44 0446421 X

 **University of Hertfordshire**

Learning and Information Services
Hatfield Campus Learning Resources Centre

College Lane Hatfield Herts AL10 9AB
Renewals: Tel 01707 284673 Mon-Fri 12 noon-8pm only

This book is in heavy demand and is due back strictly by the last date stamped below. A fine will be charged for the late return of items.

ONE WEEK LOAN

D1477200

WITHDRAWN

RUNNING WATER

More Technical Briefs on Health, Water and Sanitation

Edited by Rod Shaw

With an introduction by Ian Smout

Intermediate Technology Publications 1999

NOTE

Technical Briefs Nos 1-32 can be found in *The Worth of Water*, also available from IT Publications.

Nos 33-52 in this second volume of Technical Briefs were prepared by WEDC. Nos 53-64 were prepared by WELL.

WEDC (The Water, Engineering and Development Centre) at Loughborough University in the UK is one of the world's leading institutions concerned with education, training, research and consultancy for the planning, provision and management of physical infrastructure for development in low- and middle-income countries.

WELL is a resource centre funded by the Department for International Development (DFID) to promote environmental health and well-being in developing and transitional countries. It is managed by the London School of Hygiene & Tropical Medicine (LSHTM) and WEDC.

IT Publications, 103-105 Southampton Row, London WC1B 4HH, UK

© WEDC/IT Publications 1999

ISBN 1 85339 450 5

Printed by Russell Press Ltd., Russell House, Bulwell Lane, Basford, Nottingham NG6 0BT.

CONTENTS

	Page
Introduction	vii
33. Maintaining handpumps	1
34. Protecting springs — an alternative to spring boxes	5
35. Low-lift irrigation pumps	9
36. Ferrocement water tanks	13
37. Re-use of wastewater	17
38. Emergency sanitation for refugees	21
39. Upgrading traditional wells	25
40. Desalination	29
41. VLOM pumps	33
42. Small-scale irrigation design	37
43. Simple drilling methods	41
44. Emergency water supply	45
45. Latrine slabs and seats	49
46. Chlorination	53
47. Improving pond water	57
48. Small earth dams	61
49. Choosing an appropriate technology	65
50. Sanitary surveying	69
51. Water, sanitation and hygiene understanding	73
52. Water — quality or quantity?	77
53. Training	81
54. Emptying latrine pits	85
55. Water source selection	89
56. Buried and semi-submerged water tanks	93
57. Surface water drainage — How evaluation can improve performance	97
58. Household water treatment 1	101
59. Household water treatment 2	105
60. Water clarification using *Moringa oleifera* seed coagulant	109
61. On-plot sanitation in urban areas	113
62. Emergency water supply in cold regions	117
63. Using human waste	121
64. Wastewater treatment options	125

INTRODUCTION

RUNNING WATER is a practical book aimed at fieldworkers who plan, design and construct water and sanitation facilities for low-income communities. These facilities must be appropriate to meet the needs and demands of the women, men and children in the community — otherwise they will not use them, care for them, or pay for their operation and maintenance. But facilities must also be designed to work properly, to be easy to operate and maintain, to be long-lasting, and to be low-cost both for construction and for operation and maintenance. With these requirements in mind, each issue of IT Publications' quarterly journal *Waterlines* contains a Technical Brief on a selected topic, to help fieldworkers whether from NGOs, community-based organizations, governments, utilities or consultants.

From the introduction of the idea in the 1980s, WEDC has been responsible for preparing these Technical Briefs, with Rod Shaw managing the process and designing each 4-page insert. I am proud of this long-term association with *Waterlines* and pleased too that the Technical Briefs have proved popular with its readers. From the beginning it was thought that they would also have an enduring value as reference material. Rather than rely on readers going through years of issues of *Waterlines* to find one which addresses their particular concerns, Technical Briefs have been compiled into books. An earlier collection — *The Worth of Water* — was published by Intermediate Technology Publications in 1991. Subsequent Technical Briefs are gathered here in chronological order, together with a subject guide. The emphasis of each topic is on practical solutions for fieldworkers. Further reading is also listed so that practitioners can seek additional information. IT Publications can supply many of the books through its 'Books-by-Post' catalogue, and staff from DFID and its partners in qualifying agencies can also obtain copies through the WELL document service.

In high-level international discussions these days there is a lot of talk about creating an 'enabling environment' in the water sector. Practitioners will recognize the importance of this as they work to overcome all the constraints in their situation. But others must also recognize that this is not an end in itself, and ultimately development relies on the work of practitioners to deliver the services which people need and demand — providing support to fieldworkers should be regarded as a crucial part of the enabling environment. *Waterlines* can be seen in this context, and it is pleasing that DFID agreed to support preparation of 12 Technical Briefs through the WELL Resource Centre. WELL is commissioned to provide expertise to DFID and its partners in NGOs, developing country governments and international agencies, and this is one example.

The continuation of existing services is often a struggle but the aim must be to do more than this, improving and expanding services to reduce the huge numbers of people with inadequate water and sanitation facilities. I believe we should be identifying appropriate solutions which are viable in each circumstance and offering people a choice from these. I hope these Technical Briefs will help fieldworkers to identify and detail these viable solutions, and help communities achieve better water supply and sanitation.

Ian Smout
Leader of WEDC
Associate Director, WELL

SUBJECT GUIDE

Appropriate technology	65-68	Manganese removal	108
Arsenic removal	108	*Moringa* seeds	15-18
Bank filtration	58-59	Percussion drilling	41-42
Boiling	102	Pit latrines	49-52, 85-88
Ceramic filtration	107	Pond filters	57-60
Chemical disinfection	103	Pumps	1-4, 9-12, 35-36
Chlorination	53-56, 103	Rainwater harvesting	13-16, 93-96
Chlorine demand	53	Recycling, of wastewater	17-20
Chlorine residual	53-54	Refugee camps	21-24, 45-48
Coagulants	15-17	Reverse osmosis	29-30
Coliforms	79	Rope-and-washer pumps	10
Collection distance	78	Rotary drilling	41, 44
Community involvement	75-76, 90-92	Rotary-percussion drilling	41, 44
Community management	2	Rower pumps	9
Community mobilization	24	Sand filtration	58-59, 105-106
Cultural aspects	90-92	Sanitary surveys	69-72
Dams	61-64	Sanitation	21-24, 49-52, 73-76
Deep-well pumps	35-36	SanPlats	51
Defecation fields	21	Septic tanks	87-88
Desalination	29-32	SHTEFIE	66-68
Direct action pumps	35-36	Sludge	87-88
Disasters	21-24, 45-48,	Sludging	41, 43
Disease transmission	74, 77	Small-scale irrigation	9-12, 37-40
Disinfection	53-56, 102-104	Solar disinfection	104
Distillation	29-32	Spillways	64
Drainage	97-100	Spring protection	5-8
Drilling	41-44	Storage, of water	102
Earth dams	61-64	Straining, of water	102
Emergencies	21-24, 45-48	Suction pumps	12, 34
Evapotranspiration	38-39	Surface water	57-60, 61-64, 97-100
Ferrocement	13-16, 93-96	Training	81-84
Filtration	58-60, 105-107	Treadle pumps	11
Floods	97-100	Underground tanks	93-96
Fluoride removal	108	Vergnet diaphragm pumps	36
Groundwater	25-28	Village water supply	90-92
Hand auger drilling	41-42	VLOM	33-36
Handpumps	1-4, 33-36	Wastewater re-use	17-20
Health aspects	18-19, 57, 69-72, 73-76, 77-80	Wastewater treatment	18-19, 87
		Water clarification	15-17
Horizontal roughing filtration	60	Water demand	45
Household water treatment	101-108	Water quality	19, 70-71, 78-80,
Hygiene education	73-76	Water quantity	73, 78-80
Insect vector diseases	77	Water requirements, irrigation	38
Iron removal	107	Water sources	5-8, 46, 69-72, 89-92
Irrigation	9-12, 17-20, 37-40	Water storage	93-96
Jetting	41, 43	Water supply	45-48
Latrine emptying	85-88	Water tanks	13-16, 93-96
Latrine seats	49-52	Water treatment	15-18, 29-32, 47, 53-60, 65-68, 101-108
Latrine slabs	49-52		
Latrines	22-23, 49-52	Water-based diseases	77
Low-cost drilling	41-44	Water-washed diseases	77
Low-lift pumps	9-12	Well lining	27
Maintenance	1-4, 33-36	Wells	25-28

33. Maintaining handpumps

Handpumps can provide a permanent source of unpolluted water which is vital for a healthy developing community. A significant proportion of installations are not in optimum working order, however, and some are broken and inoperable for long periods. One of the major factors contributing to this waste of resources is inadequate or non-existent maintenance.

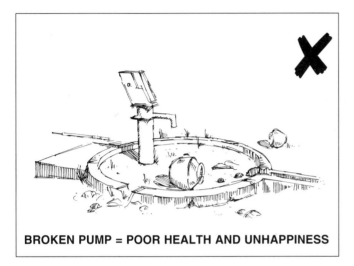

BROKEN PUMP = POOR HEALTH AND UNHAPPINESS

WORKING PUMP = HEALTH AND HAPPINESS

What is maintenance?

✘ It is **not** carrying out repairs when a pump finally breaks down.
✔ It **is** periodically inspecting an installation and replacing parts that are worn or show other signs of deterioration. Its aim is to prolong the life of the pump and avoid unexpected breakdowns.

Planned preventative maintenance is an organized system of inspections on a daily, weekly, monthly, and yearly basis, which should maximize the time for which a pump can deliver good drinking-water.

A typical schedule is given below but will vary for different pump types.

Daily:	■ Pump operation ■ Pump and base cleanliness ■ Wastewater drainage ■ Comments of users	**Weekly:**	■ Lubricate moving parts ■ Check tightness of nuts and bolts ■ Check security of pump on base
Monthly:	■ Check output rate ■ Check for condition of concrete base	**Yearly:**	■ Remove downhole assembly ■ Inspect and replace parts where necessary

Whenever remedial actions are required these are to be carried out and records kept at all stages.

This system of work requires proper scheduling, and experience has shown that the organization is just as important as the physical work.

Maintaining handpumps

The handpump is installed for the benefit of the community and it is reasonable to expect some contributio
as they will want to look after their investment. In the past, maintenance has been organized in one of

A. Community management

All inspections, repairs, renovations and replacements are carried out by members of the community or under the community's direction.

Support to the community can be provided by private enterprise (spares and services) or purchased from a government agency.

Advantages
- ✔ Fast response to problems
- ✔ In control of own affairs
- ✔ Develop pride in own abilities and achievements

Disadvantages
- ✘ Needs motivated people with appropriate level of skill
- ✘ May require engineering facilities
- ✘ Need to hold expensive stock of spares

B. Centrally managed w involvement (tiered s

Simple routine inspectio people using the pump, looking after many hand inspections, overhauls, a

Advantages
- ✔ Community retains re responsibility
- ✔ Back up for major pro
- ✔ Pride in maintaining p

Disadvantages
- ✘ Community depender
- ✘ Delays awaiting actior
- ✘ Skilled team needs to
- ✘ Expensive vehicles re

■ While the centrally managed system (C) would seem the easiest to set up, it is the least effective in the long term.
■ The compromise option (B) with both central and community involvement is perhaps the most common choice, but i
■ Totally self-sufficient communities (A) are not widespread, but with the increasing availability of Village Level Operat
■ Increasingly, private enterprise is being seen as the channel for the purchase of spares and expertise.

Typical maintenance points

Seals
- Leather or rubber seals will wear
- Leather will deteriorate rapidly if allowed to dry

Valves
- Ball valves can wear and leak from the 'water hammer'
- Rubber/leather valves will deteriorate from fatigue

'T' bar on handle reduces wear on bearings

Pivots and rub
Stuffing box -

Paint pump annually

Check pump c

Bolts and nuts - Check weekly

Concrete base
- Clean daily -

Downhole assembly
- Remove and check annually, replace worn or broken parts, and reassemble

Pump rod and

Rising main
- Corrosion or wear of threads

Cylinder - Wea

Piston valve

Piston seal/wa

When well is first constructed, surge pump to remove sand

Foot valve/che
Strainer/scree

Maintaining handpumps

continued use by the community. This is especially so if the community has initiated the pump installation,
...ways:

...mmunity
...)

...repairs are carried out by the
...entralized specialist group
... will visit periodically for major
...airs.

...e measure of control and

...eyond local resources
...developed

... on another organization
...tral group
...erly resourced to be effective

C. Centrally managed

All work is carried out by a central agency.

Advantages
- ✔ Smaller stock of spares required per pump
- ✔ Concentration of skills and resources

Disadvantages
- ✘ Slow response to remedy breakdowns
- ✘ High cost and possibly poor service
- ✘ Routine inspections may not be carried out
- ✘ No involvement or commitment by the community

...t always been sustainable.
...Maintenance (VLOM) - designed pumps, this option is becoming more feasible.

...ubricate weekly
...y
...ally

...ainage - Check daily
...water drain
...needed

...ghtness and wear

...ioration

...r deterioration

...ar or deterioration

The significance of maintenance

When handpumps are to be installed a lot of time is spent considering the type of pump and the installation details, but ease of maintenance is of vital importance as it will influence the effectiveness and life of the pump.

Ease of maintenance should influence choice as much as hydrogeology.

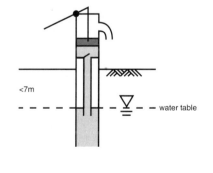
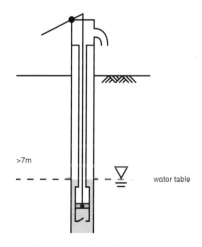

Shallow water table - Suction pump
Easy maintenance as all parts are above ground

Deep water table - Lift pump
Complex maintenance as some operative parts are below ground

Maintaining handpumps

Community involvement

No matter what system of management is adopted, user involvement is vital for the long-term effectiveness of the handpump. The best way to achieve this is by the appointment of a *pump caretaker* who, after proper training and the supply of a tool kit, will carry out the following duties:

The pump caretaker's duties:
- To carry out inspections daily, weekly, monthly
- To keep records of all checks and work
- To monitor pump output rate
- To keep pump and base clean and clear of refuse
- To train people how to use the pump properly
- To make simple repairs or replacements
- To request help for major problems
- To keep a supply of spare parts
- To ensure surplus water is drained away
- To give guidance in health care

The pump caretaker should:
- Be female (if culturally or socially acceptable)
- Be aged 18-35
- Live close to the pump
- Be physically fit and active
- Be acceptable to the community
- Be a pump user
- Have own means of support as the pump caretaker's job is only part-time
- Be self-motivated

In order to emphasize their responsibility, pump caretakers should receive payment.

Physical maintenance

There are so many types of handpump that it is not within the scope of this Technical Brief to detail specific maintenance points, but it should be noted that whenever parts rub or rotate, wear will take place. Lubrication will minimize the wear, but routine inspections will confirm the rate of deterioration and decide when a replacement is required.

All nuts and bolts should be kept tight, as excessive play encourages wear.

Each type of pump will require a different level of maintenance, and one reason for keeping records is to review procedures to check if they are appropriate.

Every handpump should be available for work 100 per cent of the time, but this cannot be achieved by only responding to breakdowns. A strategy of Planned Preventative Maintenance will keep the non-working time to a minimum.

Further reading

Colin, J., *VLOM for Rural Water Supply: Lessons from experience,* WELL, London, 1999. (http://www.lboro.ac.uk/well)
Kennedy, W.K. and Rogers, T.A., *Human and animal-powered water-lifting devices,* IT Publications, London, 1985.
Pacey, A., *Handpump maintenance in the context of community well projects,* IT Publications, London, 1977.
UNICEF, *India Mark-II handpump installation and maintenance manual.*
World Bank / Rural Water Supply Handpumps Project, *Community water supply — the handpump option,* World Bank, 1986.

Prepared by Bob Elson, Richard Franceys and Rod Shaw

WEDC Loughborough University Leicestershire LE11 3TU UK
www.lboro.ac.uk/departments/cv/wedc/   wedc@lboro.ac.uk

34. Protecting springs — an alternative to spring boxes

Spring boxes
Many publications describe the protection of springs from contamination using spring boxes similar to the one in Figure 1.

The construction of such a structure takes time and money, and in many cases it may not be really necessary.

Why use a spring box?
A spring box can be useful as:
- **A sedimentation chamber** where particles of sand carried in the spring water can settle out.
- **A storage chamber,** which is useful for springs where the peak rate of demand exceeds the rate of flow of the spring.
- A method of **protecting the spring water** from contamination.
- A way of **collecting the spring water** by giving it an easy flow path from the aquifer into a delivery pipe.

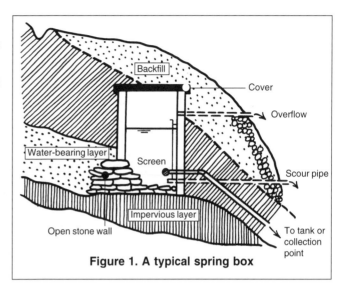

Figure 1. A typical spring box

So it follows that:
if...   water from a spring can be **protected and collected** without a spring box,
and if... **no sedimentation is needed** because the water carries only a low level of suspended solids,
and if... **no storage is needed** because the water flows at a rate sufficient to meet the peak demand...
then no spring box is required.

Advantages of protecting a spring without using a spring box
- **Simpler design**   ○ easier for local masons to copy   ○ quicker construction   ○ less cement and local materials needed
- **Reduced cost**   ○ more communities may be able to afford to protect a spring
                         ○ protection of low flowing springs becomes more financially viable to funding agencies
- **Suits flatter sites** ○ no provision is needed for the depth of the storage/sedimentation chamber

Each spring is unique and all designs need adapting to suit the type of spring and the topography. The following 7-step guide to protecting a spring without using a spring box assumes a gravity spring with a flow rate of up to about 35 litres/minute, flowing from a granular unconfined aquifer lying on a fairly impermeable strata. Use of the design with higher flow rates is possible in some instances, but more than one delivery pipe and a larger concreted sump behind the headwall are recommended for such flows.

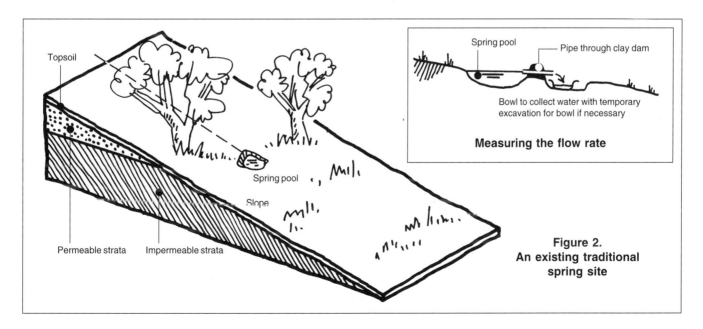

Figure 2. An existing traditional spring site

Protecting springs — an alternative to spring boxes

1. Choose the site

First, check the following:
- Does the community want the spring protected?
- Are local construction materials available?
- Is the community willing to contribute by way of organization, money, labour, materials, transport, etc.?
- Check the existing spring flow rate (e.g. by inserting a pipe into a clay dam at the overflow point and by recording the time taken to fill a container of known volume — see Figure 2). By protecting the spring you may be able to achieve an increased flow, but consider whether there is likely to be enough water to meet the demand.
- During what season of the year are you measuring the flow? Has the community noticed a change of flow with season?
- Latrines, animal pens, etc. uphill of the spring pose a pollution risk. They should be at least 50m away.
- The site needs to slope sufficiently to dispose of surface water and wastewater.

2. Find the main flow path/s

- Clear the site of bushes, long grass, etc.
- Starting at the highest point(s) at which there is evidence of water issuing from the soil, excavate narrow trenches uphill following the direction from which most of the water is flowing. Stop when the trench is about 1.0m deep if sufficient water is flowing into the end of the trench from the 'eye' of the spring. If there is more than one main source then several trenches can be joined.
- If there are no main flow paths it may be necessary to excavate a seepage trench across the slope, to intercept water seeping through the aquifer (see Figure 3).

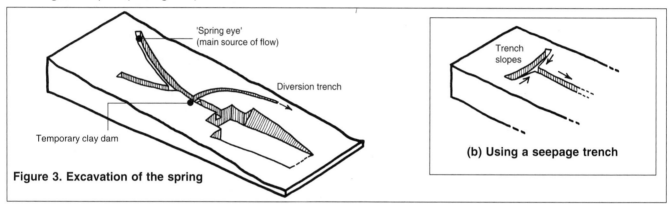

Figure 3. Excavation of the spring

(b) Using a seepage trench

3. Protect the source and choose how to convey the water

The spring eye at the head of the trench should be surrounded with clean stones through which water can flow into the trench. Stones of between 10 and 40mm diameter are usually suitable for this, but larger ones can be used. A layer of stones about 100mm deep will usually suffice, and this should then be covered with a layer of rocks and a layer of 'puddled clay' about 100mm deep. This clay is prepared by wetting and kneading it underfoot until it is uniformly plastic. Its purpose is to prevent surface water and grains of backfill material from entering the stone-filled channel. (See Figure 4.)

Once the clay has been trodden into place the remainder of the trench can be backfilled. The excavated material can be used, and it should be compacted by foot in layers of about 100mm. The final layer in the trench should be of topsoil, which is raised a little above the ground to compensate for the future settlement of the backfill. This topsoil should be planted with creeping grass plants to prevent soil erosion.

It is useful for future reference to measure and record the positions of the spring eyes from some permanent features, such as the corners of the headwall, so that if problems occur the eyes can be found again quickly. Alternatively, a large flat stone can be embedded in the topsoil on the surface above the eye to mark its position.

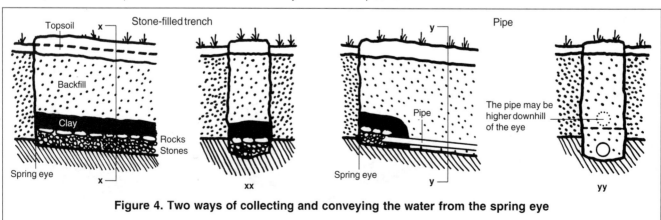

Figure 4. Two ways of collecting and conveying the water from the spring eye

Protecting springs — an alternative to spring boxes

There are two options for conveying the water from the protected source to the headwall:

- **A stone-filled trench** If the bottom of the trench that you have dug to the eye is in fairly impermeable strata, and if it is smooth and sloping, then it can be backfilled with a layer of stones in a similar way to that already explained and shown in Figure 4. Water can then flow between the stones and along the trench to the headwall.
- **A pipe** If it is affordable, a 30-50mm internal-diameter plastic pipe can be used to carry the water. This has three advantages:
  - it eliminates the water losses that can occur from stone-filled trenches;
  - it protects the spring water from pollution as it travels to the headwall; and
  - if desired, and if the topography is suitable, the pipe can allow a delivery point to be above ground. This means that a large excavation for the headwall and for the wastewater drain is avoided, and only a small headwall structure is needed to support the delivery pipe.

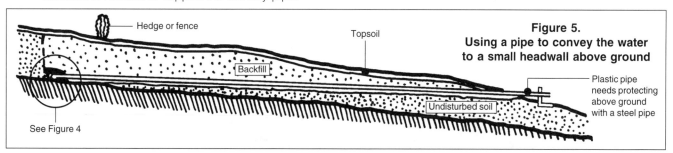

Figure 5. Using a pipe to convey the water to a small headwall above ground

4. Choose the position of the headwall

In choosing the position of the headwall consider the following:
- If stone-filled trenches are being used to convey the water, then the closer the headwall is to the spring eye the less chance there is of the spring water being polluted in the trench, or of it being lost from the trench; **but**
- if the excavation for the headwall is too close to the spring eye, it may adversely affect the local water-flow pattern, and the spring might be lost; **and**
- the closer the headwall is to the spring eye, the higher and stronger the wall will have to be. A smaller headwall further away from the eye may be the most economical solution if the cost of the increased length of pipe or stone-filled trenches is not excessive.

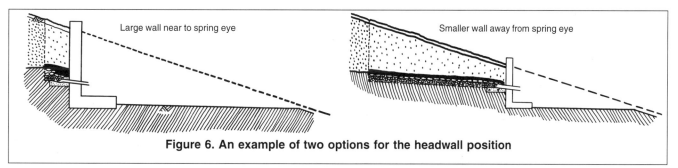

Figure 6. An example of two options for the headwall position

5. Excavate for and construct the headwall, wingwalls and access

1: Divert the water

The headwall needs to be constructed in dry conditions. During construction the flow of spring water into the excavation can be prevented in two ways:
- By using a temporary clay dam in the main trench to divert the water into a diversion trench (see Figure 3).
- By carrying the water over the excavation in a pipe or locally made gutter (e.g. of bamboo or split banana tree trunk) supported on forked sticks (see Figure 7).

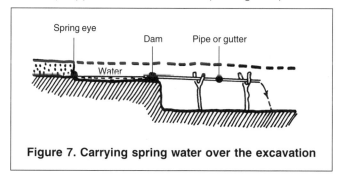

Figure 7. Carrying spring water over the excavation

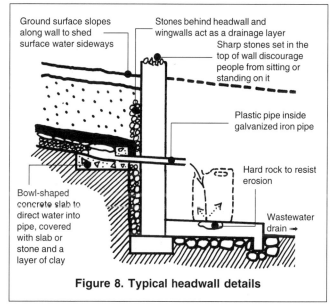

Figure 8. Typical headwall details

Protecting springs — an alternative to spring boxes

2: Build the headwall and wingwalls
A variety of materials can be used to build the headwall, but well-burnt bricks, concrete blocks, or stones laid in a 1:3 cement:sand mortar are the most common materials. Small wingwalls at right angles to the headwall help to support it against the soil pressure. A drainage layer behind the headwall prevents hydrostatic forces from acting on the wall.

3: Build the apron slab and steps
The purpose of the apron slab and steps is to give users convenient access to the delivery pipe. The apron slab also protects the wall foundations and it should slope to discharge wastewater into the drain. If the users approach from both sides of the spring, then two sets of steps may be provided, or a simple footbridge can be built across the drain. It may be possible to avoid the use of steps altogether by providing access alongside the drain serving the apron slab.

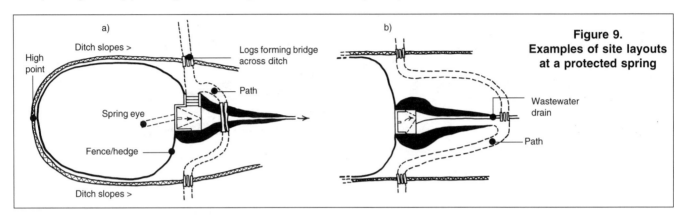

Figure 9. Examples of site layouts at a protected spring

6. Construct the surface water diversion ditch and fence
The area immediately above and uphill of the spring eyes needs to be fenced off to prevent pollution from people or animals. The fence should extend at least 10m uphill of the spring eyes. To prevent polluted surface water from flowing through the fence onto the site of the spring, a free-draining ditch should be constructed uphill of the fence. The alignment of the ditch needs to be chosen to suit the topography before the fence is built beside it. A hedge of animal-resistant bushes usually makes a good permanent fence. The area inside the fence should be planted with creeping perennial grass if it is available, and other vegetation should be cut down to keep the area tidy and to prevent roots from penetrating the spring water trenches.

7. Train the community
The training of the community should start even before work commences on the spring. They need to understand how common water-borne diseases are spread, and how the clean water from the spring can easily become polluted before they drink it. Regular visits to the spring by a health educator are usually necessary to encourage proper water use and spring maintenance. Where the community appoints a spring committee and one or two caretakers, these people can be trained to be aware of the health risks associated with the spring and they can be encouraged to explain these to the users.

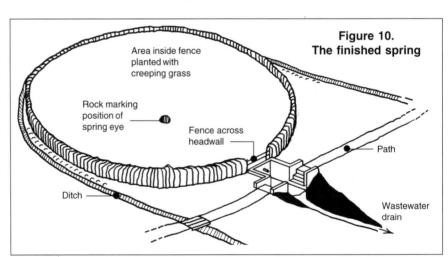

Figure 10. The finished spring

Further reading
Hofkes, E.H. (ed.), *Small Community Water Supplies: Technology of Small Water Supply Systems in Developing Countries*, IRC Technical Paper No. 18, Wiley & Sons, Chichester, 1983.

Rous, T., 'Protecting a shallow seepage spring', *Waterlines*, Vol.4 No.2, IT Publications, London, 1985.

Tobin, V., and Cairncross, S. Technical Brief No. 3: Protecting a Spring, *Waterlines*, Vol.3 No.3, IT Publications, London, 1985.

Prepared by Brian Skinner and Rod Shaw

WEDC Loughborough University Leicestershire LE11 3TU UK
www.lboro.ac.uk/departments/cv/wedc/   wedc@lboro.ac.uk

35. Low-lift irrigation pumps

Small-scale irrigation is a key factor in the development of many rural communities in developing countries. A variety of water-lifting devices have been used in the past and this Technical Brief describes some of the most practical and efficient pumps in use today.

An irrigation pump needs to be able to deliver a large volume of water over a long period of time, and in order to achieve this the available human power needs to be used efficiently. Arms and shoulders are normally used to operate machinery, but higher power outputs can be achieved by using the whole body. A pedalling action gives the most power, as it uses the leg muscles, the largest in the body.

The water being pumped can come from a variety of sources – wells, boreholes, streams, rivers, and ponds.

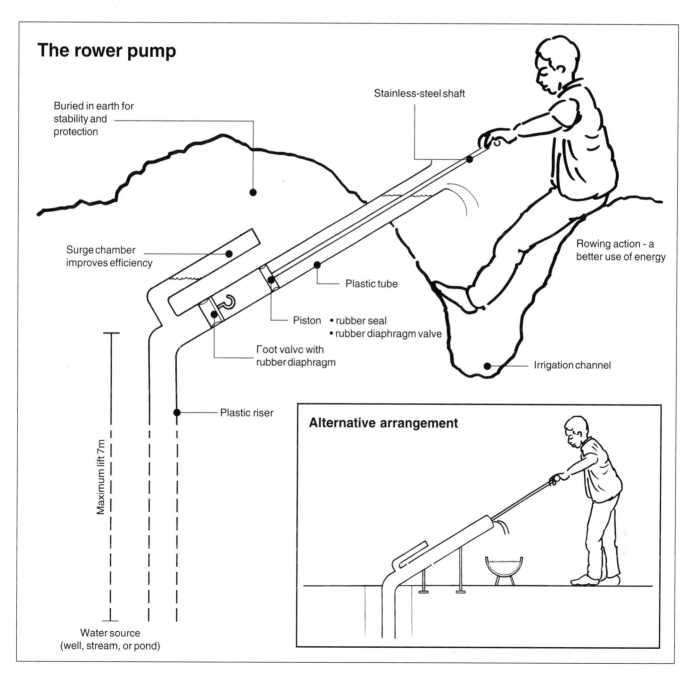

Low-lift irrigation pumps

The rope-and-washer pump

Based on a principle developed in ancient China, the recent design is a VLOM pump that can be built from materials that are available in most communities.

The main pulley is an old car-tyre rim and this is used to turn a rope knotted to hold a series of rubber washers, made from car tyres. The riser pipe can be either plastic or bored-out bamboo.

Water can be lifted from up to 20m below the pump and delivered 5m above it, with output rates of up to 50 litres per minute depending on lift.

Sources can be wells, large diameter boreholes, ponds, and streams.

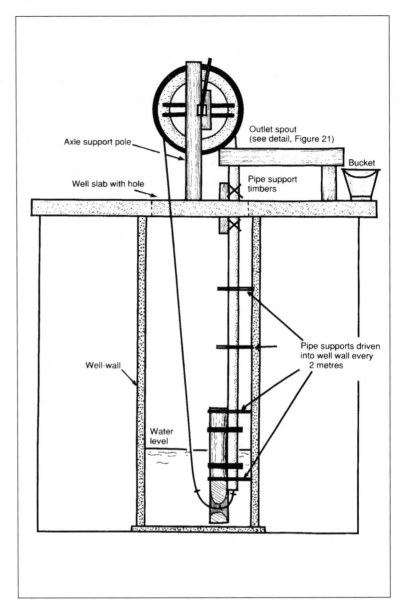

Alternative arrangements

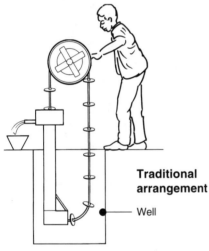

Traditional arrangement

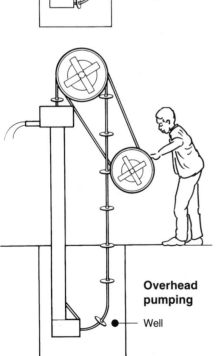

Overhead pumping

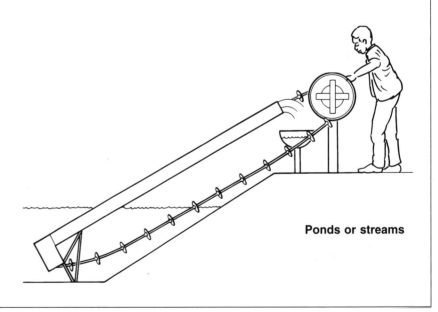

Ponds or streams

Low-lift irrigation pumps

The treadle pump

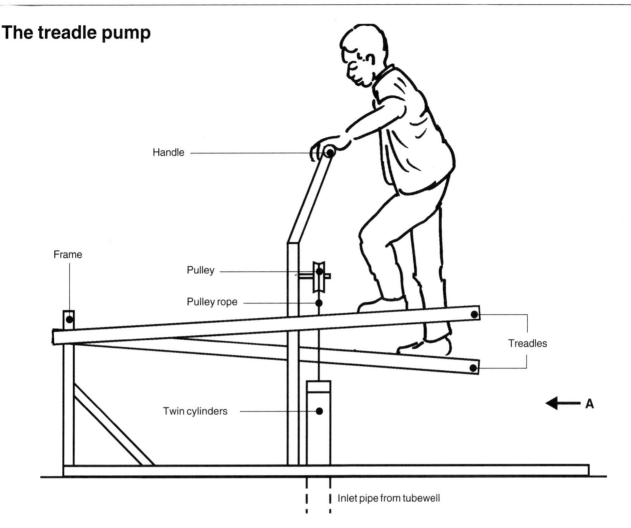

Because the treadle pump is operated by the most powerful muscles of the human body, it can be operated for longer than other human-powered pumps.

The pump originally developed in Bangladesh has since been modified so that the discharge is pressurized. In this form it is capable of lifting water up to 20m above the pump, but because it works on a suction principle it can only be a maximum of 6m above the source. Water can be pumped from wells, boreholes, streams, and ponds, and discharge rates of 50 litres per minute can be achieved. This is a VLOM (Village Level Operation and Maintenance) pump but welding facilities are required for its manufacture.

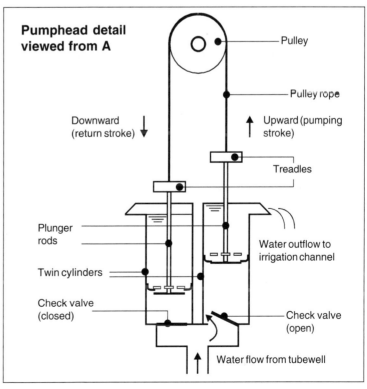

Pumphead detail viewed from A

Low-lift irrigation pumps

Suction pumps

Suction pumps exist in a great variety of designs. They can be used for low-lift irrigation purposes, but they have several disadvantages:

- lower discharge rates

- need to be fixed in a stable position over a well or borehole

- less efficient for prolonged use as arm and shoulder muscles used

- more sophisticated manufacture required

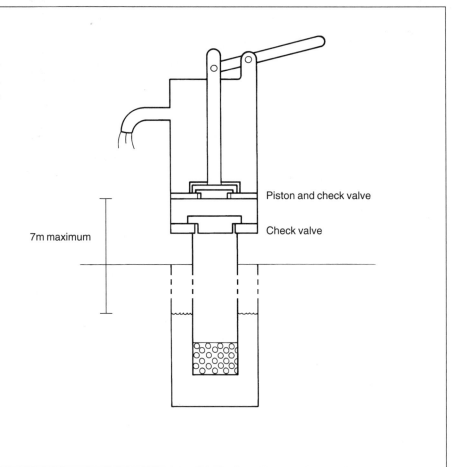

Alternative methods

There are other ways of lifting water for irrigation that have been in use for many generations. These include the Egyptian *shadoof*, the *dhone* and various water wheels and devices such as the *mohte*, which use animal power. Wind power can also be used to power irrigation devices.

Further reading

Barnes, G., *The development of a manual irrigation device - the twin treadle pump,* The International Conference on Agricultural Equipment for Developing Countries, 1985.
Brelenburg, C. and Allen, H., *How to Make and Use the Treadle Irrigation Pump,* IT Publications, 1995.
Fraenkel, P., *Water Pumping Devices,* IT Publications, London, 1997
Lambert, R.A., *How to Make a Rope and Washer Pump,* IT Publications, London, 1990.
Lambert, R.A. and Faulkner, R.D., *Manually operated irrigation pumps,* 4th African Water Technology Conference, 1989.
Kennedy, W.K. and Rogers T.A., *Human and Animal-Powered Water-Lifting Devices,* IT Publications, London, 1985.
Stern, P., *Small-Scale Irrigation,* IT Publications/IIIC, London, 1979.
World Bank/UNDP, *Community water supply - the handpump option,* World Bank, 1986.

Prepared by Bob Elson and Rod Shaw

WEDC Loughborough University Leicestershire LE11 3TU UK
www.lboro.ac.uk/departments/cv/wedc/   wedc@lboro.ac.uk

36. Ferrocement water tanks

Ferrocement consists of a cement-rich mortar reinforced with layers of wire mesh, sometimes with additional plain wire reinforcement for added strength. Tanks made of ferrocement are used in many countries for the collection and storage of water for drinking, washing, for animal use and irrigation.

Ferrocement tanks have several advantages over tanks made of concrete or brick:

- They are usually cheaper to build and require less skilled labour.
- They are able to withstand shock better, as ferrocement is more flexible.
- Smaller ferrocement tanks are portable.

Plastering the inside of a ferrocement tank

Ferrocement tanks vary in capacity, size, and shape. They are built by hand-trowelling layers of cement mortar onto a wire frame which is either free-standing or held in place by temporary or permanent structures known as 'formwork'.

Ferrocement is only needed for tanks of capacities greater than 1000 litres. Below this size, cement mortar alone is strong enough to withstand the applied loads.

Tanks used for storing drinking-water must always be covered to avoid contamination and so maintain drinking-water quality.

Fittings are usually built into the ferrocement during construction. These include:

- one or more taps for water collection;
- a drainage tap (or wash-out) at the bottom of the tank, to be used when cleaning;
- an inlet pipe; and
- an overflow pipe. This must be screened to prevent insect entry.

A tank may be sited above ground or below ground, or it may be partially sunk (provided its base is situated well above groundwater level).

Methods of construction

There are different methods of constructing ferrocement water tanks.

- **Building tanks without using formwork**
  This method requires a stiff wire frame around which flexible mesh such as 'chicken wire' is wrapped. The first layer of mortar is applied by a mason working on one side of the tank, with an assistant on the other side holding a plastering float in the right place to allow the mortar to be compacted without it falling through the mesh.

- **Building tanks using temporary formwork**
  Temporary formwork, made of wood, flat or corrugated sheets of steel, or coiled pipe is placed against one face of the tank during the application of the initial layers of mortar. The formwork is removed before plastering the inside of the tank.

- **Building tanks using permanent formwork**
  Formwork, such as corrugated sheets, may be left in place permanently, plastered on both the inside and the outside.

- **Centrally produced tanks**
  Smaller ferrocement tanks can be centrally produced and transported in one piece to their point of use. Larger tanks have to be built in sections. Central production has the advantage of better quality control, but the transportation costs may make the tanks unaffordable.

Ferrocement water tanks

Simplified stages of construction of a ferrocement tank

1. The foundation is prepared

- The levelled site is cleared of debris
- Topsoil is removed to a depth of 100mm

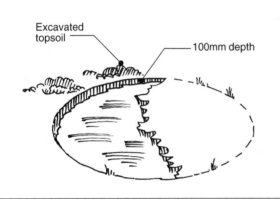

2. The base slab is laid

- The site is covered with 50mm of concrete
- Wire reinforcement, with steel supports attached, is placed over the concrete before it sets
- A second layer of concrete is laid over the wire reinforcement to ground level

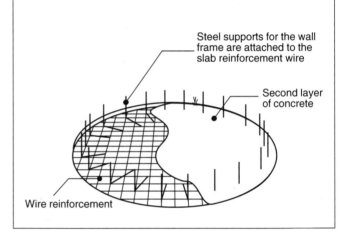

3. The wall frame is constructed

- The wall reinforcement is attached to the steel supports using binding wire
- For larger tanks, wooden shuttering may be constructed to give added support to the wire frame

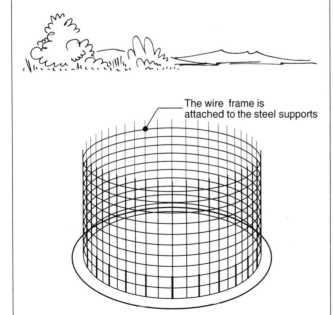

4. Mesh is attached to the frame

- Layers of wire mesh (or 'chicken wire') are attached to cover the frame on both the outside and the inside

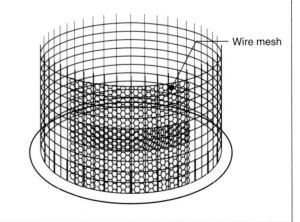

Ferrocement water tanks

5. Fittings are installed

- Fittings are attached to the wire reinforcement before the walls are plastered. The tap below is held securely in place by a plate welded to the pipe and embedded in the ferrocement.
- Additional cement mortar is plastered around the tap to prevent leakage.

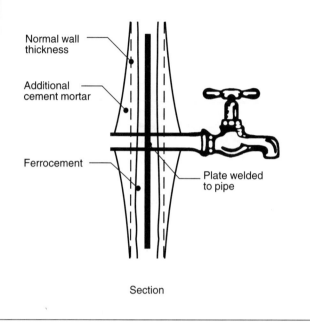

Section

6. Plastering

- The tank is plastered on the outside first

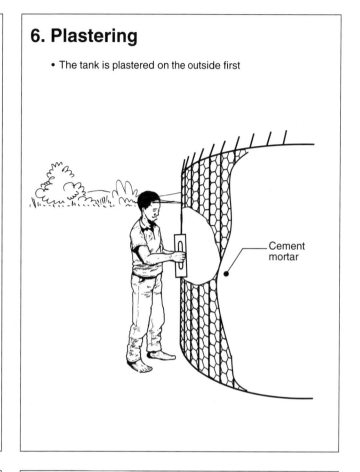

7. The plastered tank

- When the cement mortar is set on the outside, stepladders are used to access the inside of the tank, which is also then plastered.

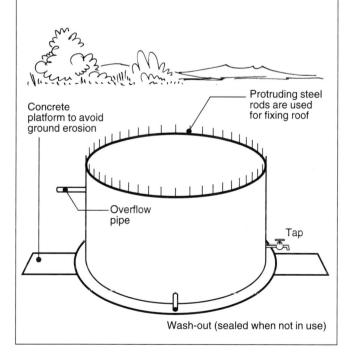

8. Roofs

- To prevent evaporation, pollution, and the breeding of mosquitoes, all tanks should have roofs.
- Arched ferrocement roofs are increasingly being used, because they are cheaper than those made of reinforced concrete.
- Roofs with spans of up to 2.5m are usually free spanning from the wall of the tank. Larger spans are usually supported by a central prop.

Roof shapes - sectional view

Ferrocement water tanks

Further points to consider

Careful selection of materials and proper mixing and curing of the cement mortar are important in order to produce watertight tanks.

The reinforcing wire

It is important to get a good distribution of wire mesh throughout the cement mortar to limit cracking.

Wire meshes come in a wide variety of types, diameters, stiffnesses, and strengths. For large tanks, welded meshes made from the larger diameters of wire (above 4mm) welded at 100mm to 200mm spacing to a square or rectangular grid are often used to provide a strong reinforcing frame, which is then covered with smaller sizes of mesh or netting.

Wire netting

Wire netting, often called 'chicken wire', is very useful for distributing loads through the mortar and into the larger diameter reinforcing wires. It is easily moulded to form spherically shaped surfaces, such as arched roofs to tanks.

The cement mortar

Cement mortar is a mixture of sand, cement, and water. The strength of the mortar depends on these raw materials, the mix ratio, and the workmanship of those who mix and use the mortar.

Sand

Clean, dry sand should be used. It should be well-graded, comprising particles of different sizes.

Cement

Cement should have been recently manufactured and have been protected from water vapour during storage and transport.

Water

The water used in the mix needs to be clean, preferably of drinking-water quality.

The cement:sand ratio

The usual ratio of cement to dry sand is 1:3 by volume. To achieve the desired ratio, a bucket can be used to accurately measure out the proportions of sand and cement. (Note that when sand from a stockpile is damp it has a greater volume than when it is dry). As cement 'bulks', it is preferable to use a full bag of known volume.

The water:cement ratio

The ratio of water to cement has an important effect on the final strength of the mortar. A ratio of about 0.4:1 to 0.5:1 (ratio of water:cement by weight) is ideal, which is equivalent to between 20 to 25 litres of water to each 50kg bag of cement.

Mixing

It is preferable to use a concrete mixer. Where this is not possible, mix the right ratios of sand and cement on a hard, clean surface until the mixture is of uniform colour. Cast a mixing slab if necessary, or use a portable mixing trough to prevent loss of cement and to prevent soil contaminating the mortar.

Add only sufficient water to make the mortar 'workable'. If the mortar is too stiff because too little water has been added it will be hard to compact; it will have a poor bond to the reinforcement; and it will not be held in place by adhesion to the formwork. If the mortar is too wet it will produce a weak and permeable tank. Water should never be visible in the mixed mortar, even when it is left undisturbed in a pile.

Mixed mortar should be used immediately. Extra water should never be added to soften the mortar once it has started to set. Cover or shade the mixed mortar in hot weather and turn the pile over regularly.

Compact the mortar well by pushing it hard against the formwork.

Curing

Once the mortar has set, keep it damp for at least two weeks and preferably longer. This curing is important for the proper gaining of strength and the prevention of cracking. It can be assisted by wetting the surfaces and covering them with polythene sheeting or wet sacking. It will still be necessary to periodically wet the surfaces before they can be allowed to dry.

Further reading

Chindaprasirt, P., et al. *A Study and Development of Low-Cost Rainwater Tanks*, Khon Kaen University, Bangkok, 1986.
Hasse, R., *Rainwater Reservoirs above Ground Structures for Roof Catchment*, Vieweg & Sohn, Germany, 1989. (Also available from IT Publications, London.)
Nissen-Petersen, E., *How to Build an Underground Tank with Dome*, ASAC Consultants Ltd., Kitui, Kenya, 1992.
Nissen-Petersen, K., *How to Build Smaller Water Tanks and Jars*, ASAC Consultants Ltd., Kitui, Kenya, 1992.
Sharma, P.C. and Gopalaratnam, V.S., *Ferrocement Water Tanks*, International Ferrocement Information Center (IFIC), Asian Institute of Technology, Bangkok, 1980.
Watt, S.B., *Ferrocement Water Tanks and their Construction*, IT Publications, London, 1978.

Prepared by Brian Skinner, Bob Reed and Rod Shaw

WEDC Loughborough University Leicestershire LE11 3TU UK
www.lboro.ac.uk/departments/cv/wedc/   wedc@lboro.ac.uk

37. Re-use of wastewater

In many arid and semi-arid countries, wastewater is becoming an increasingly important source of irrigation water. The demands of growing urban communities for both food and water require the agricultural sector not only to increase food production but also to reduce its use of natural water resources. At the same time the volume of sewage effluent is increasing, and safe disposal can be difficult. The use of reclaimed wastewater for irrigation is the obvious solution, but few people have expertise in the full range of technology involved.

This Technical Brief considers situations where it may be appropriate to re-use wastewater for agriculture and introduces the different types of wastewater re-use scheme. It also provides a recommended guide to water quality for irrigation, and outlines, using diagrams, the necessary procedures for treating wastewater.

When to re-use wastewater

There are several questions to consider:

■ **What are the water requirements of the community?**

Many communities in most developing countries do not have reliable access to supplies of clean water. As the demand for water increases, making more efficient use of water becomes more important. Water re-use should be seriously considered before water availability is matched by water demand (Figure 1). Note that not all water needs to be treated to potable standards. Most wastewater re-use is informal and goes largely unrecognized by the public and by many professionals.

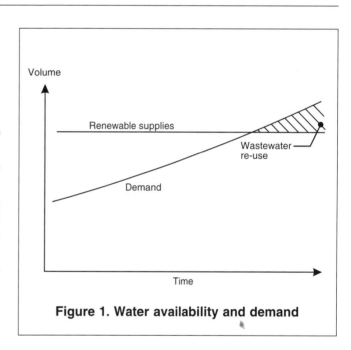

Figure 1. Water availability and demand

■ **Is the content of the wastewater harmful?**

Wastewater may contain chemicals which are harmful to the growth and development of plants. It may also contain bacteria and other organisms which are harmful to agricultural workers and to those who handle, cook, or eat the plants. Wastewater may even contain bacteria and other organisms which, when eaten by animals, may in turn infect the people who eat the contaminated meat. Figure 2 examines the health risks in relation to the level of contamination and the re-used wastewater control measures.

■ **What will the wastewater be used for?**

It is important to first consider which water uses are the major ones, and efforts should then be made to be more economical in these sectors. Industry and agriculture require large volumes of water, but the quality need not always be high. Water demands for irrigated agriculture are considerable. (For example, since 1949 agricultural water consumption in Israel has ranged from 71.3 to 83.3 per cent of the total water consumption.)

■ **Is it economical to re-use wastewater?**

The costs of treating the wastewater adequately as opposed to using conventional water resources should be carefully considered and the more economical option chosen.

Re-use of wastewater

Control measures	Wastewater or excreta	Field or pond	Crop		Worker	Consumer
	Level of contamination				Level of risk	
No protective measures	High	High	High		High	High
Crop restriction	High	High	High		High	Safe
Application measures	High	High	Safe		Safe	Safe
Human exposure control	High	High	High	**Desirable sanitary barrier**	Low	Low
Partial treatment in ponds	Low	Low	Low		Safe	Low
Partial treatment by conventional methods	Low	Low	Low		Low	Low
Partial treatment in ponds, plus crop restriction	Low	Low	Low		Safe	Safe
Partial treatment by conventional methods, plus crop restriction	Low	Low	Low		Low	Safe
Partial treatment, plus human exposure control	Low	Low	Low		Safe	Low
Crop restriction, plus human exposure control	High	High	High		Low	Safe
Full treatment	Safe	Safe	Safe		Safe	Safe

Figure 2. Wastewater re-use: Control methods and health risks
(Adapted from WHO Technical Report 778. Health guidelines for use of wastewater in agriculture and aquaculture.)

Types of wastewater re-use schemes

Wastewater re-use may be 'direct' or 'indirect'.

Direct re-use is the planned and deliberate use of treated wastewater for some beneficial purpose, including drinking.

Direct potable re-use is not popular and is limited to a few places including Windhoek in Namibia and Denver in the United States. It is generally unacceptable to the public because of both the expense and the attitudes of the community. Studies have shown that people will drink wastewater from an indirect source unless there is evidence to suggest that it is unsafe to do so. People will not, however, drink water from a direct source unless it is proven to be safe.

Indirect re-use refers to water that is taken from a river, lake, or aquifer which has received sewage or sewage effluent.

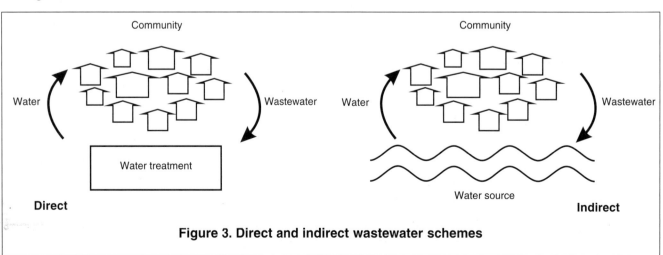

Figure 3. Direct and indirect wastewater schemes

Re-use of wastewater

Category	Re-use conditions	Exposed group	Intestinal nematodes[b] (arithmetic mean no. of eggs per litre[c])	Faecal coliforms (geometric mean no. per 100 ml[c])	Wastewater treatment expected to achieve the required microbiological quality
A	Irrigation of crops likely to be eaten uncooked, sports fields, public parks[d]	Workers, consumers, public	≤1	≤1000[d]	A series of stabilization ponds designed to achieve the microbiological quality indicated, or equivalent treatment
B	Irrigation of cereal crops, industrial crops, fodder crops, pasture, and trees[e]	Workers	≤1	No standard recommended	Retention in stabilization ponds for 8 to 10 days or equivalent helminth and faecal coliform removal
C	Localized irrigation of crops in category B if exposure to workers and the public does not occur	None	Not applicable	Not applicable	Pre-treatment as required by the irrigation technology, but not less than primary sedimentation

[a] In specific cases, local epidemiological, sociocultural, and environmental factors should be taken into account, and the guidelines modified accordingly.
[b] *Ascaris* and *Trichuris* species and hookworms.
[c] During the irrigation period.
[d] A more stringent guideline (≤200 faecal coliforms per 100 ml) is appropriate for public lawns, such as hotel lawns, with which the public may come into direct contact.
[e] In the case of fruit trees, irrigation should cease two weeks before fruit is picked, and no fruit should be picked off the ground. Sprinkler irrigation should not be used.

(Adapted from WHO Technical Report 778. Health guidelines for use of wastewater in agriculture and aquaculture.)

Figure 4. Recommended quality of water for irrigation[a]

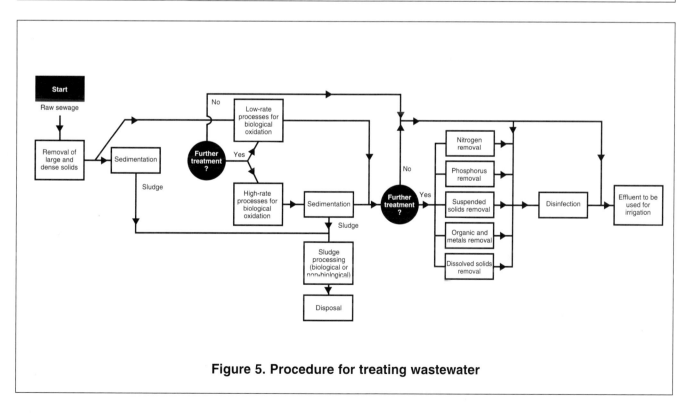

Figure 5. Procedure for treating wastewater

Re-use of wastewater

Further points to consider

- Studies in South America, Asia, and the Middle East have shown that farmers prefer to grow produce in the following order of priority:

1. Vegetables (to earn a regular income);
2. Fruit (to earn a regular income or foreign exchange);
3. Cereal crops (of lower value);
4. Fodder crops (of low value);
5. Other crops for which there is a demand (herbs, spices, flowers, etc.)

- The re-use of wastewater for irrigation has been most successful near cities, where wastewater is easily available and where there is a market for agricultural produce.

- The storage of treated wastewater may be necessary, because supply may not match demand (Figure 6).

- Re-use requires:
  o careful planning;
  o adequate and suitable treatment;
  o careful monitoring;
  o appropriate legislation; and
  o the implementation of legislation and quality standards.

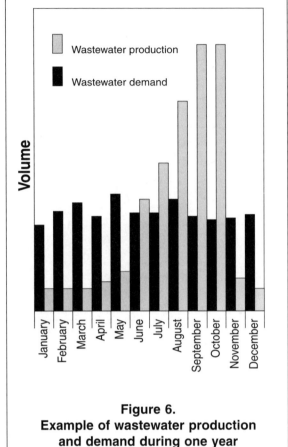

Figure 6.
Example of wastewater production and demand during one year

Conclusions

- Re-use can help to maximize the use of limited water resources.
- Wastewater re-use can contribute to national development.
- Environmental damage caused by re-use should be minimized.
- Health risks associated with re-use should be minimized.
- Collaboration between users, authorities, and the public is needed.
- Exchange of experiences is very important.
- Government support and encouragement is needed.

Further reading

Mara, D., and Cairncross, S., *Guidelines for the Safe Use of Wastewater and Excreta in Agriculture and Aquaculture,* WHO, Geneva, 1989.
Pescod, M.B. *Wastewater Treatment and Use in Agriculture,* FAO Irrigation and Drainage Paper 47, Food and Agriculture Organization, Rome, 1992.
Shuval, H.I., Adin, A., Fattal, B., Rawitz, E., and Yekutiel, P., *Wastewater Irrigation in Developing Countries: Health Effects and Technical Solutions,* World Bank Technical Paper No. 51, World Bank, Washington, 1986.
World Health Organization, *Health Guidelines for the Use of Wastewater in Agriculture and Aquaculture,* Report of WHO Scientific Group, Technical Report Series No. 778. Geneva, 1989.

Prepared by Michael Smith and Rod Shaw

WEDC Loughborough University Leicestershire LE11 3TU UK
www.lboro.ac.uk/departments/cv/wedc/   wedc@lboro.ac.uk

38. Emergency sanitation for refugees

The immediate provision of clean water supplies and sanitation facilities in refugee camps is essential to the health, well-being and, in some cases, even the survival of the refugees. Sanitation is usually allocated a much lower priority than clean water, but it is just as important in the control of many of the most common diseases found in refugee camps.

Sanitation is the efficient disposal of excreta, urine, refuse, and sullage. As indiscriminate defecation is normally the initial health hazard in refugee camps, this Technical Brief outlines ways in which it can be controlled temporarily while long-term solutions are devised.

Immediate measures

The technical options for emergency excreta disposal are limited and simple. If they are to work, however, they must be managed well and **be understood and supported by the community.** The immediate tasks at a new camp include:

- obtaining the services of a good translator and consulting with all interested parties including representatives of the refugees, aid agencies, and government officials;
- surveying the site to gather information on existing sanitation facilities (if any), the site layout, population clusters, topography, ground conditions, and available construction materials;
- preventing defecation in areas likely to contaminate the food chain or water supplies; and
- selecting areas where defecation may safely be allowed.

Preventing defecation in certain areas

When a large group of people are excreting indiscriminately, it is necessary, first of all, to protect the food-chain and water supplies from contamination. This means preventing people defecating on:

- the banks of rivers, streams, or ponds which may be used as a water source. If water is to be abstracted from shallow wells, then it is important to ensure that these wells are situated upstream of the defecation areas; or
- agricultural land planted with crops, particularly if the crops are soon to be handled or harvested for human consumption.

Keeping people away from such areas may not be easy, particularly where traditional habits make such practices common. It may be necessary to construct a physical barrier, such as a fence, which may need patrolling. Immediate measures to control indiscriminate defecation should not be solely negative, though; it is much better to designate areas where defecation is allowed than to fence off those that are not.

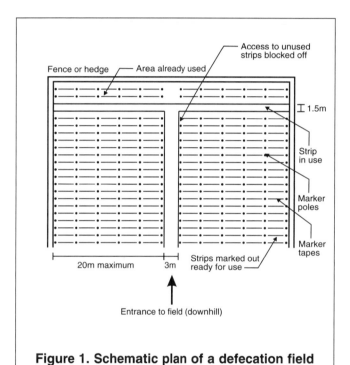

Figure 1. Schematic plan of a defecation field

Defecation fields

Areas with fixed boundaries within which defecation is permitted are known as 'excretion' or 'defecation' fields. The use of these fields localizes pollution, and makes the management and the cleaning of the site easier. They should be located carefully so that they are easily reached by the community but do not pollute water supplies or sources of food. It is better if there are a number of fields at roughly equal intervals over the site area, as this will reduce the walking distance for most users and allow for flexibility of operation and the separation of the sexes.

The defecation field should be as large as possible, but it should not be open for use all at once. It is better to divide the field into strips so that a different strip can be used each day. The area of the field farthest from the community should be used first, so that people do not have to walk across contaminated ground to reach the designated area (Figure 1).

21

Emergency sanitation for refugees

Intermediate measures

The life-span of the excretion fields is not long because the areas polluted by excreta cannot be used again unless a system is established to cover the excreta with soil. Their purpose is to allow time for latrines to be built.

The ideal solution is to provide each family with their own latrine, but unless this is the simplest of structures (Figure 2), it is neither feasible nor advisable immediately. In the early days it will not be known how long it will be before the situation which has caused the disruption to the refugee community will return to normal. Furthermore, refugees will naturally be unsettled at this stage, and may be unable or unwilling to commit themselves to the maintenance of permanent or semi-permanent structures that may suggest that their displacement will last a long time.

An intermediate solution is required. It is usual for this to be some form of communal latrine, as communal latrines are quick and cheap to construct. Some are commercially available, but these are expensive and take time to transport to the site. In most cases, 'trench' latrines provide the simplest solution (Figure 3).

In many cases, families provided with the basic materials and tools can build their own latrines. They should be encouraged both to keep their latrines clean and to construct a new one before the old one fills up.

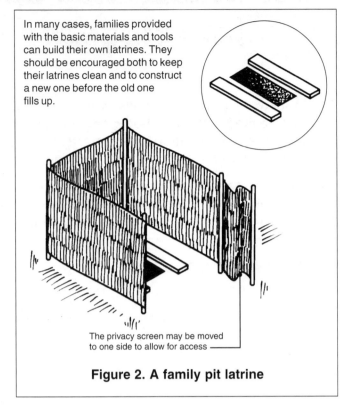

The privacy screen may be moved to one side to allow for access

Figure 2. A family pit latrine

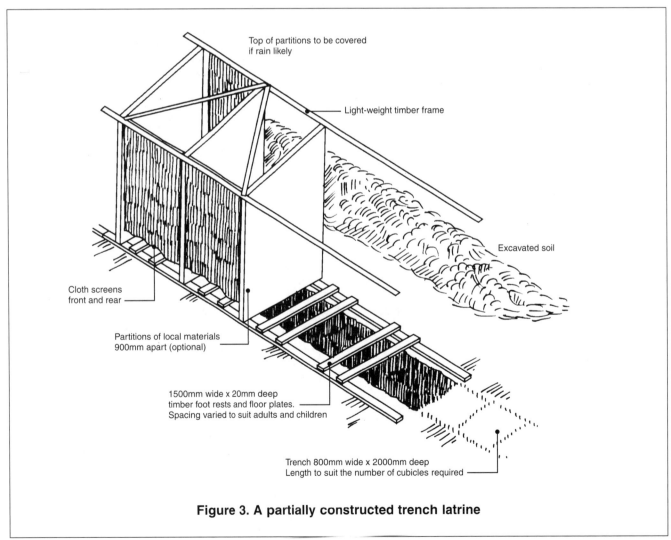

Top of partitions to be covered if rain likely

Light-weight timber frame

Excavated soil

Cloth screens front and rear

Partitions of local materials 900mm apart (optional)

1500mm wide x 20mm deep timber foot rests and floor plates. Spacing varied to suit adults and children

Trench 800mm wide x 2000mm deep Length to suit the number of cubicles required

Figure 3. A partially constructed trench latrine

Emergency sanitation for refugees

Trench latrines

A trench latrine is a rectangular hole in the ground. The hole should be dug as deep as possible — about 2m and may be lined with timber where there is danger of collapse. It may be of any convenient length, usually between 5 and 10m, and between 1 and 1.5m wide. The trench is spanned by pairs of wooden boards on which the users squat (see Figure 3). There is a gap between the boards through which the users excrete. Preferably, each pair of boards is separated by a simple screen to provide privacy. In wet weather a roof is needed to prevent the trench from filling up with rainwater. A drainage ditch should be built to divert surface water.

Each week the contents of the trench are covered by a 100 to 150mm-deep layer of soil. This will reduce the smell and prevent flies from breeding in the trench. When the bottom of the trench has risen to within 300mm of the surface, the trench is filled in and the latrine is closed.

A trench latrine system is very labour-intensive and requires constant supervision. Not only must the contents of each latrine be covered each day, but new latrines must be prepared, old ones filled in, and regularly used latrines cleaned. Close supervision is essential. A poorly maintained latrine will quickly become offensive to the community and will not be used.

Making use of existing facilities

If refugees settle in or near urban areas, it may be possible to make use of existing facilities such as sewers, public toilets, bucket latrines, or stormwater drains.

Mobile package latrines

In the North, mobile package latrines are common. There is no reason why they cannot be used in other places provided provision is made for the ultimate disposal of the excreta.

Borehole latrines

In areas with deep soil, many borehole latrines can be built in a short time using hand augers. The holes are usually 30 to 50cm in diameter and 2 to 5m deep. The top of each hole is lined with a pipe, and two pieces of wood comprise the footrests. Borehole latrines should be closed when the contents are only 500mm from the surface (Figure 4).

Long-term solutions

Trench or borehole latrines are only an intermediate solution because their operation is so labour-intensive and requires constant supervision. As soon as it becomes obvious that the community is likely to remain disrupted for any length of time, longer-term solutions should be sought. In most cases, some form of on-site sanitation will be most appropriate.

(a) Drilling the borehole using a hand auger

(b) The borehole and footrests

Figure 4. A borehole latrine

Emergency sanitation for refugees

Community mobilization

The safe disposal of excreta in refugee camps is primarily the result of good supervision and management, and this can only be achieved with the full co-operation of the community. It is essential, therefore, that the community is fully consulted at all times and that their views are considered and their suggestions implemented. Problems may arise as immediate sanitation measures usually conflict with personal habits and social customs, but strict control measures at the outset, when people are still disorientated, will usually help them to become accustomed to new ideas and methods. Later, the supervision of the excretion fields and the policing of protected areas can easily be done by the community itself.

The co-operation of the community will only be gained and retained if it is kept fully informed of what is being done and why. Information is communicated best through group meetings and personal contact.

Group meetings

Group meetings can be used to advise the community about what is proposed, how the systems will operate, and why they are important (Figure 5). Such meetings should give the community an opportunity to question and advise on what is being proposed. It is important that every effort is made to include as many of their views as possible. In the early stages, the community is usually too tired and confused to contribute to the proposals, but this stage quickly passes and soon the community will start to take a lively interest in its surroundings.

Individual contact

Group meetings are effective at passing on general information, but there is a possibility that some sections of the community will not be reached and these meetings are not appropriate for dealing with individual problems. For these situations, personal contact is more appropriate. Improving hygiene awareness, particularly among mothers, is usually better achieved on a one-to-one basis or within very small groups. Such education is long term and slow, but it should be started as soon as possible since it is often easier to establish new behaviour patterns in a community before it becomes established.

Labour

The day-to-day operation of latrines and programmes of education require substantial labour. While key management posts are likely to be provided from outside the area, much of the initial routine work can be done by the community. In most cases the community is only too willing to help since it gives people something to do, prestige, and possibly a source of income. Latrine supervision is not a popular job and will almost certainly have to be paid for. Motivation may be improved by providing a uniform and protective clothing or installing special bathing facilities for supervisors. People working on latrine operation require little or no training; those involved in health education and information dissemination will require more.

Figure 5. A group meeting

Prepared by Bob Reed and Rod Shaw

WEDC Loughborough University Leicestershire LE11 3TU UK
www.lboro.ac.uk/departments/cv/wedc/   wedc@lboro.ac.uk

39. Upgrading traditional wells

Wells have been used to obtain water since ancient times. Some wells have been in continuous use for hundreds of years. Others are fairly new, but have been built by traditional methods. Good quality water can usually be obtained from a well that is properly constructed, maintained, and used. Some traditional wells are excellent. Others are not and need upgrading.

Water drawn from surface sources – streams, rivers, lakes, and tanks – is often polluted. Groundwater, however, is filtered naturally as it passes through the soil to reach the *aquifer,* where it is stored underground. The word 'aquifer' means 'water-carrying'. Aquifers occur in pores, voids and fissures. Pores are the spaces between the mineral grains. Some aquifers carry water in layers of sand or gravel, others in fractured rock. Some are only a few metres thick, others hundreds of metres. Some underlie a few hectares, others extend across thousands of square kilometres. Some are hundreds of metres down and others lie just under the surface. Most traditional wells take water from shallow aquifers.

Where groundwater is deep, hand or motorized pumps may be necessary.

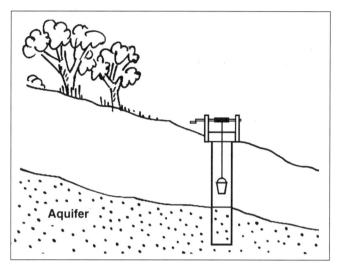

Many traditional wells are unsatisfactory because ...

1.

They are near pit latrines, rubbish dumps, or animal pens, which pollute the groundwater that is the source for the well.

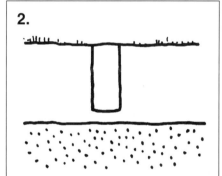

2.

They are not deep enough, so water is inaccessible in dry weather when the groundwater level is low.

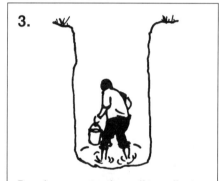

3.

People may enter the well to collect water, making it dirty and increasing the risk of spreading guineaworm disease in areas where it is endemic.

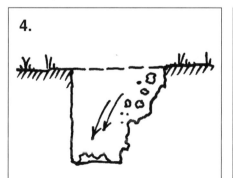

4.

The sides collapse, reducing the amount of water that can be collected and allowing surface water to get into the well from the top.

5.

The top is open, so polluted water and dirt falls in, polluting the well water, and there is a risk of shoes, animals, and children falling in.

6.

Dirty buckets and ropes are used for drawing water, thereby polluting the well.

Upgrading traditional wells

How can wells be upgraded?

1.
Move rubbish dumps and animal pens far away from the well.

A minimum distance of 25 metres is recommended, and the dump or pen should not be on higher ground than the well.

A well which is very near to a deep pit latrine should not be upgraded.

A new well should be dug at least 25 metres from the pit, and the old well closed off.

2.
Deepen the well so that the bottom reaches the groundwater during dry weather

At the end of the dry season, when the groundwater is likely to be at its lowest level, use two buckets to deepen the well.

Continue to remove soil and water when groundwater is reached. It is often possible to dig two metres or so into the aquifer, although this depends on the rate of recharge. Obviously work has to stop when the water is filling the well faster than it can be removed.

An experienced well digger should be chosen to dig into the aquifer, because of the danger of collapse.

A windlass can be used to lower and raise people working in the well, and to raise soil and water.

3.
Improve the well so that people and animals cannot get in

This can be done by lining the well as illustrated opposite and by providing a cover as illustrated below.

A well should be situated at least 25m away from latrines, rubbish dumps and animal pens

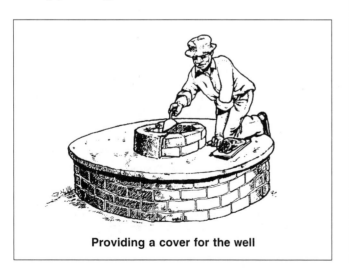

Providing a cover for the well

Deepening the well

Upgrading traditional wells

4.
Line the well

Materials that are often used for lining wells include:

- **Bricks:** Kiln-dried bricks should be used at the bottom (below the highest level of groundwater) and at the top.
- **Natural stone** (masonry): Untrimmed rough stone is particularly suitable in the aquifer.
- **Concrete blocks:** These may be curved to suit the shape of the well, or ordinary rectangular blocks can be used.
- **Concrete rings:** These may be made on site, or may be cast in a central yard and brought to the well.
- ***In-situ* concrete:** Concrete is cast around steel shutters in the well. This is an expensive method and is not normally used for upgrading wells.

The bottom of the well should be lined so that water can flow in:

- Concrete rings may have holes made during casting, or holes may be carefully broken through the concrete after casting. Gaps may be formed by putting small stones between the rings.
- Masonry linings may be built as a 'dry wall' with no mortar between the stones.
- Joints between bricks or blocks should have the smallest amount of mortar necessary to hold them in position. Sometimes no mortar is put in vertical joints. Alternatively, unjointed masonry is used as a base for bricks or blocks.

The remainder of the lining should be back-filled with soil that is 'tamped' down by treading or pushing down with poles.

If the soil around the well is very loose it may be necessary to back-fill with weak concrete.

The top part of the lining should have good mortar joints and should extend at least 300mm above ground level. The back-filling should be impervious, using 'puddled' clay or weak concrete.

Puddled clay is clay which has been thoroughly mixed with water — it is 'kneaded' in the way that dough is kneaded when making bread.

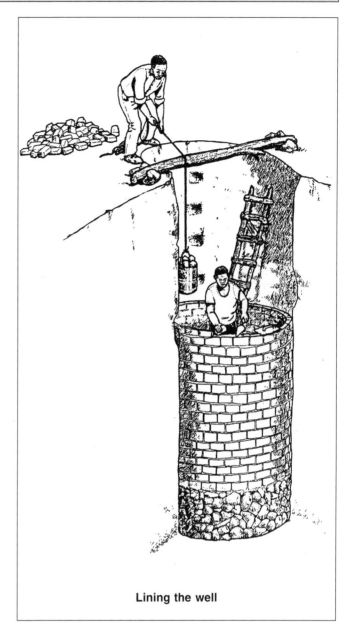

Lining the well

5.
Provide a cover or a parapet wall

A cover can be a reinforced concrete slab with a hole for buckets to pass through. It can be cast near the well and lifted to fit on the lining.

Parapet walls are extensions of the wall lining about a metre above the top of the apron.

Putting a cover in place

Upgrading traditional wells

6.
Install a windlass or handpump

A windlass enables buckets full of water to be raised easily by turning a handle.

- Stout wooden posts or steel pipes are embedded about 600mm deep in concrete on each side of the well.
- A windlass may be purchased ready-made or can be fabricated from 20 - 25mm steel pipe or rod.

Handpumps should be firmly fixed to the cover.

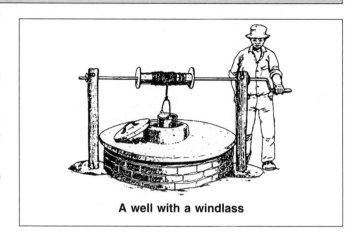

A well with a windlass

What else needs to be done?

Even if a traditional well has been improved by giving attention to all the matters listed above, it may still be unsatisfactory without other improvements.

A well head which is surrounded by a pool of spilled water is unpleasant for users, can lead to pollution of the water in the well, and may provide a breeding place for mosquitoes.

Therefore...
an apron and a drain should be provided.

The apron should be wide enough for people to stand on. It should be made of concrete that is between 75 and 100mm thick, preferably reinforced by steel rods or mesh. It should have a smooth surface sloping down to the drain and a rim to prevent water going over the edge.

The drain should be a cement-lined channel, at least four metres long, that takes spilled water to a soakage pit, an animal-watering trough, or a vegetable garden.

This Technical Brief has been adapted from *Rural water supplies and sanitation* by Peter Morgan, The Blair Research Laboratory, Ministry of Health, Harare, Zimbabwe. Published by Macmillan, London, 1990.

Adapted for IT Publications by John Pickford   Additional drawings by Rod Shaw

WEDC Loughborough University Leicestershire LE11 3TU UK
www.lboro.ac.uk/departments/cv/wedc/   wedc@lboro.ac.uk

40. Desalination

Desalination describes a range of processes which are used to reduce the amount of dissolved solids in water. Dissolved solids are often referred to as total dissolved solids (TDS), and are measured in mg/l. As a means of producing potable water, desalination is usually an expensive option. It is often associated with electricity generation plants, from which both electricity and waste heat are available. This Technical Brief outlines the processes and techniques involved, but also presents low-cost methods of desalination by distillation.

Desalination is used to produce potable water from water sources containing dissolved chemicals, and is most often used when water sources are salty; producing fresh water from sea water or brackish water. Natural waters may be classified approximately according to their TDS values:

Type of water	TDS value (mg/l)
Sweet waters	0-1000
Brackish waters	1000-5000
Moderately saline waters	5000-10 000
Severely saline waters	10 000-30 000
Seawater	More than 30 000

The main application of desalination techniques is the production of fresh water on ships, islands, and in the coastal regions of some very arid Middle Eastern countries. The water that is produced may be so pure that consumers do not like the lack of taste, and small quantities of salt water may then be added to improve the flavour.

There are several methods of water desalination. The most appropriate method can be selected on the basis of the TDS value of the raw water.

Process	TDS value (mg/l)
Ion exchange (not described here)	500-1000
Electrodialysis (not described here)	500-3000
Reverse osmosis (standard membranes)	500-5000
Reverse osmosis (high-resistance membranes)	Over 5000
Distillation	Over 30 000

Of the desalination methods available, the two main ones are:

- **reverse osmosis;** and
- **distillation followed by condensation.**

Reverse osmosis

Osmosis is a technique which plants use to absorb water from the soil and to transport the water up the stem to all parts of the plant. Dilute and more concentrated solutions are separated by a semi-permeable membrane, which acts like a very fine filter. The semi-permeable membrane allows water molecules to pass, but prevents the movement of salt or other dissolved chemical molecules.

If two saline solutions (or water and a saline solution) are separated only by a semi-permeable membrane, there will be a transfer of water through the membrane to the more concentrated saline solution. The passage of water will continue until a stable condition is reached, with the difference of liquid levels across the semi-permeable membrane being referred to as the osmotic pressure. The osmotic pressure varies with temperature and the concentrations of the two solutions (Figure 1a).

By applying pressure (in excess of the osmotic pressure) to the salt-water solution, the process can be reversed, and water molecules from the salt-water solution can be forced through to the other side of the semi-permeable membrane (Figure 1b).

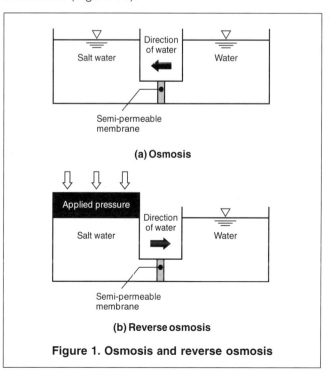

Figure 1. Osmosis and reverse osmosis

Desalination

The flow of water through the semi-permeable membrane is very slow, so a large area of membrane is needed. The membrane is easily torn, and needs to be supported carefully. Membranes are frequently wrapped into a spiral, or formed into bundles of tubes which are sealed at one end (Figure 2). Reverse osmosis installations require further refinements in order to prevent damage or blockage and to operate successfully. The salt water needs to be filtered first to remove particles which might damage the membranes, and chemical additives may be needed to control the pH and to minimize the deposition of salt on the membrane surface (Figure 3).

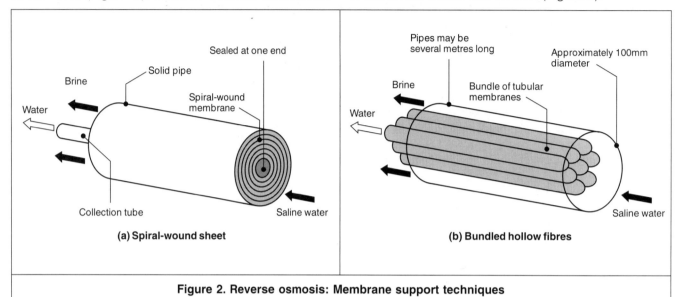

Figure 2. Reverse osmosis: Membrane support techniques

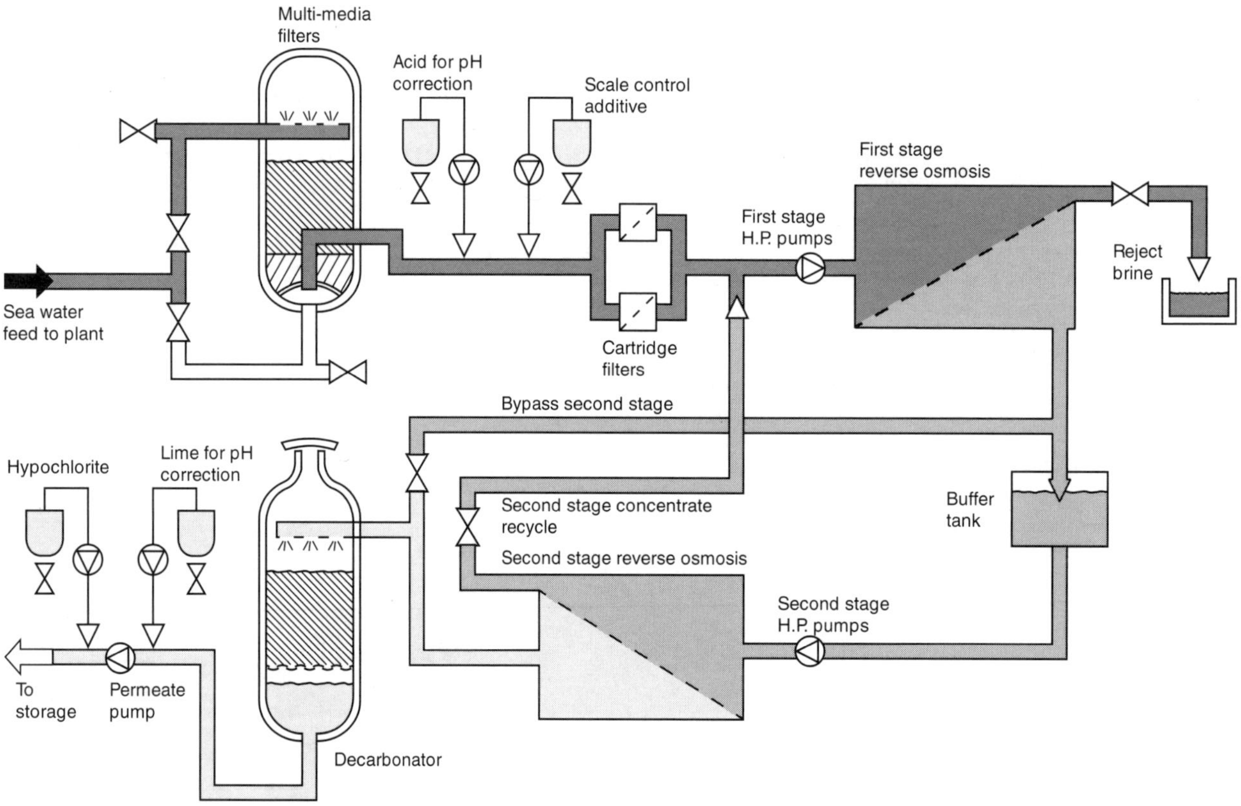

Figure 3. Schematic diagram of a reverse osmosis plant for sea water

Desalination

Distillation

The collection of water by distillation and condensation is a survival technique which can be used to collect small quantities of water from the ground. A hole is dug in the ground, and a cup or bowl is placed in the bottom of the hole. A sheet of plastic is stretched across the hole, its edges are weighted with soil to hold it in place, and a small stone is placed in the centre. Water evaporates from the soil, condenses on the underside of the plastic sheet, and collects in the cup or bowl (Figure 4). Solar energy can be used to evaporate water from salt water for household or community water supplies by constructing sealed units covered with glass (Figure 5). There are problems with these units: growth of algae on the underside of the glass sheet must be controlled, and the unit must be effectively sealed.

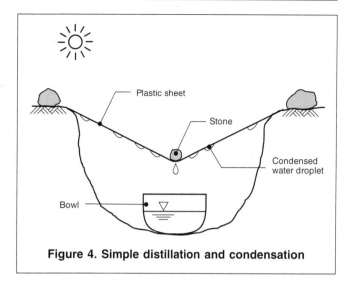

Figure 4. Simple distillation and condensation

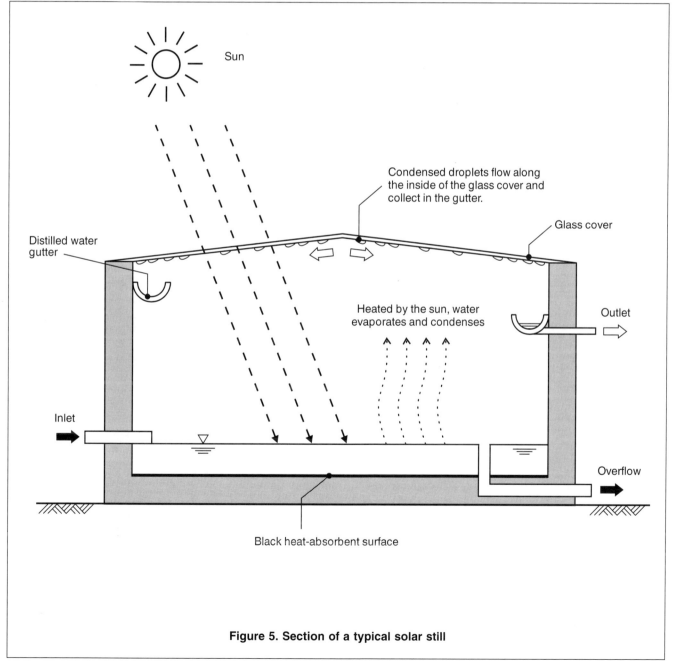

Figure 5. Section of a typical solar still

Desalination

Alternative low-cost distillation method

A simple, low-cost method to desalinate sea water by distillation is used in some countries where fuel is available. It requires basic kitchen utensils: two pots, one four times the size of the other, and a plastic sheet. The smaller pot is placed inside the larger one and weighed down with a stone.

Sea water is poured into the outer container up to the brim of the inner one. The larger pot is sealed using a plastic sheet and a piece of string so that the plastic sheet sags in the middle (Figure 6). This home-made still is then placed on any heat source such as a stove or wood fire, at low temperature. In a few minutes the sea water in the outer container starts to evaporate. As the plastic sheet prevents the steam from escaping, the droplets condense into the smaller vessel. Residual salt remains in the outer pot.

To conserve fuel, a still can be placed on top of a cooking-pot which is used everyday, such as a rice pot. As it boils or simmers, the 'waste' heat is usefully harnessed.

Care should be taken to ensure that all pots are stable and out of reach of young children.

Large-scale distillation

Large-scale distillation units use a process known as *Multi-stage flash distillation*. This conserves energy, but the equipment used is expensive and sophisticated. Multi-stage flash distillation units operate on the principle that by raising the pressure in a sealed container, the boiling point of a liquid can be raised; and that by lowering the pressure in a sealed container the boiling point of a liquid can be reduced.

Steam is used to heat a saline solution, which is driven through a series of 15 to 40 evaporation chambers. Each evaporation chamber is at a lower temperature and pressure than the preceding one, and some water vapour is evaporated or 'flashed off' from each chamber as the temperatures and pressures decrease. The hot salt water or 'brine' can cause corrosion or an accumulation of chemical deposits, and chemical additives are needed to control this.

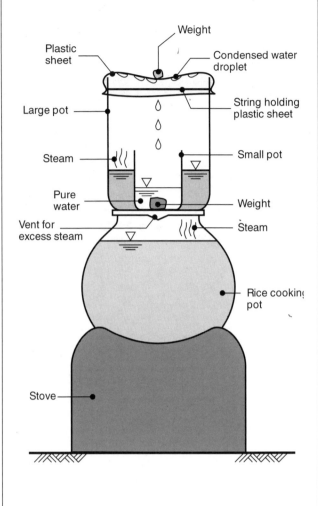

**Figure 6.
A simple, low-cost method of desalination by distillation and condensation**

Reproduced from RYFO Handout 655,
(courtesy of Dr Felix Ryan)

Further reading

Porteous, A. (Ed.), 'Desalination technology, developments and practice', *Applied Science,* London, 1983.
Tiwari, G.N., *Advance Solar Distillation Systems,* Kamala Kuteer Publications, India, 1985.
United Nations, *Solar Distillation as a Means of Meeting Small-Scale Water Demands.* Publication E.70.II.B.1., New York, 1970.
United Nations, *Water Desalination in Developing Countries.* Publication 64.II.B.5, New York, 1964.
Yates, R., Woto, T. and Tlhage, J.T., *Solar-Powered Desalination: A case study from Botswana,* IDRC, Ottawa, 1990.

Prepared by Michael Smith and Rod Shaw

WEDC Loughborough University Leicestershire LE11 3TU UK
www.lboro.ac.uk/departments/cv/wedc/   wedc@lboro.ac.uk

41. VLOM pumps

What is a VLOM pump?

A **VLOM** pump is one which can be operated and sustained using **V**illage **L**evel **O**peration and **M**aintenance.

The term **VLOMM** is also used, meaning **V**illage **L**evel **O**peration and **M**anagement of **M**aintenance.

This addition emphasizes the role of users as the *managers of maintenance* – they may choose to use someone from *outside* the village to assist with more complicated repairs. Not all maintenance and repair needs to be done by the villagers for a pump to be classed as a VLOM pump.

Why are VLOM pumps needed?

Many handpump projects have failed because of:

- the absence of a sustainable system of handpump maintenance and repair;
- the installation of pumps which were not suitable for the heavy usage they received;
- the use of pump components which were damaged by corrosive groundwater; and
- a lack of community involvement in important aspects of the project planning.

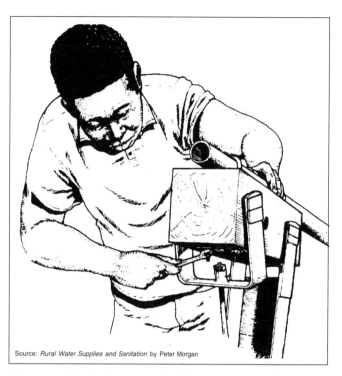

Source: *Rural Water Supplies and Sanitation* by Peter Morgan

The careful choice of a VLOM handpump can help solve the first three of these problems, but unless the community is involved from the beginning in the planning of the pump project and the management of the maintenance, it is unlikely that the handpump will be sustainable.

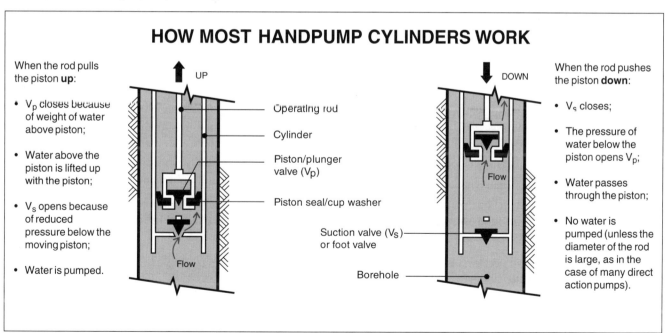

HOW MOST HANDPUMP CYLINDERS WORK

When the rod pulls the piston **up**:
- V_p closes because of weight of water above piston;
- Water above the piston is lifted up with the piston;
- V_s opens because of reduced pressure below the moving piston;
- Water is pumped.

When the rod pushes the piston **down**:
- V_s closes;
- The pressure of water below the piston opens V_p;
- Water passes through the piston;
- No water is pumped (unless the diameter of the rod is large, as in the case of many direct action pumps).

In most handpump cylinders a piston is alternately raised and lowered by a rod (or a string of rods joined together) which is connected to a handle, or sometimes to a flywheel and crank. These pumps are called **reciprocating handpumps.** The figure above illustrates how most cylinders work.

VLOM pumps

There are three types of reciprocating handpump.

Type of pump	Maximum pumping lift (m)	Cylinder above or below groundwater
Suction	7 - 8.5	Above
Direct action	15 - 25	Below
Deep-well	45 - 80+	Below

One of the basic aims of a VLOM handpump is to make all the main wearing parts easy to reach and replace, and to reduce the wear and tear on the pump by good design. The main wearing parts of a reciprocating handpump are:

- The piston seal, which rubs against the inside face of the cylinder.
- The piston valve and suction valve (or foot valve), which are constantly opening and closing.
- The bearings in the pump-head, which are subjected to constantly changing loads.

Suction pumps

Traditional design

Rod hanger bearing
(Note: Some pumps have a third bearing and mechanism to eliminate sideways movement of the rod)
Handle bearing
Piston, cup seal, and piston valve
Suction valve
Cylinder (often of cast-iron)
Ground level (concrete)
Suction pipe (installed in borehole casing or directly in the ground)

VLOM designs similar to traditional pumps are available, but often with these improvements:

- better suction valves to eliminate priming;
- smoother cylinder walls to reduce wear on piston seals;
- wear-resistant seal instead of leather (e.g. nitrile rubber); and
- better bearings to prevent the pivot pins wearing out the cast-iron (e.g. using hardened bushes around the pivot pins).

Rower design

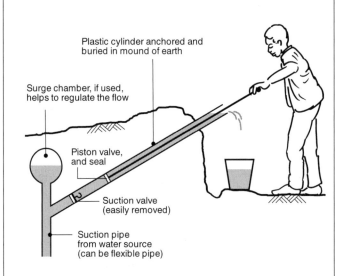

Plastic cylinder anchored and buried in mound of earth
Surge chamber, if used, helps to regulate the flow
Piston valve, and seal
Suction valve (easily removed)
Suction pipe from water source (can be flexible pipe)

The rower pump has other VLOM features.

- It allows very easy access to the piston and suction valve.
- It is relatively cheap and easy to manufacture.
- On some versions, the valves can be replaced using discs cut from car inner tubes.

Important: If this pump is to be used to draw drinking-water, care must be taken to avoid contaminating the cylinder, for example by using poor-quality priming water.

The cylinder of a suction pump is usually above ground level.

Main advantages

- Easy access to wearing parts because they are usually all above the ground.
- Fast delivery of water because of the large piston diameter (traditional designs), or long piston stroke (rower design).

Main disadvantages

- Only suitable for pumping lifts of up to about 7m.
- May need to be 'primed' by adding water to the cylinder if the suction valve leaks overnight.
- Villagers will often use polluted water to prime the pump, thereby contaminating it.
- Pump designs are often not suitable for use by more than about 50 people per day unless frequent repairs and replacements are carried out.

Direct action pumps

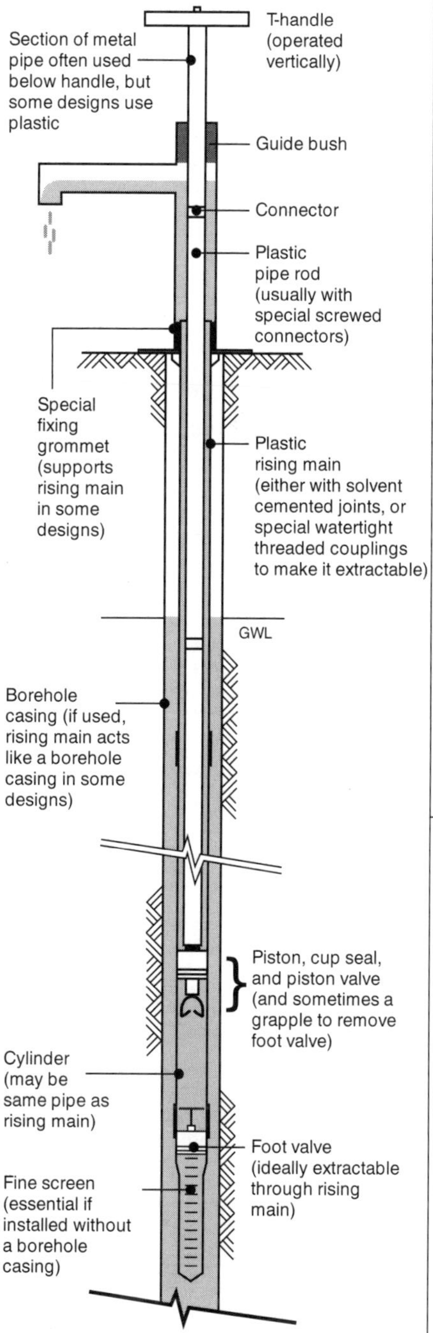
T-handle (operated vertically)
Section of metal pipe often used below handle, but some designs use plastic
Guide bush
Connector
Plastic pipe rod (usually with special screwed connectors)
Special fixing grommet (supports rising main in some designs)
Plastic rising main (either with solvent cemented joints, or special watertight threaded couplings to make it extractable)
GWL
Borehole casing (if used, rising main acts like a borehole casing in some designs)
Piston, cup seal, and piston valve (and sometimes a grapple to remove foot valve)
Cylinder (may be same pipe as rising main)
Foot valve (ideally extractable through rising main)
Fine screen (essential if installed without a borehole casing)

In most direct action handpump designs, the piston is raised and lowered by a 'T' bar handle, which is directly connected to an air-filled plastic pipe 'rod'. This rod floats in the water in the rising main, reducing the force needed on the upstroke. On the down-stroke, as more of the pipe rod enters the water in the rising main, it displaces an equal volume of water, so the pump delivers water on both the up-stroke and the down-stroke.

Deep-well pumps

Traditional design

Pump-head: Most pump-head lever handles work on a similar principle to the handle shown for the traditional suction pump. Some pumps use just one pivot and a chain (or belt) and quadrant system, such as in the India Mk II, shown below.

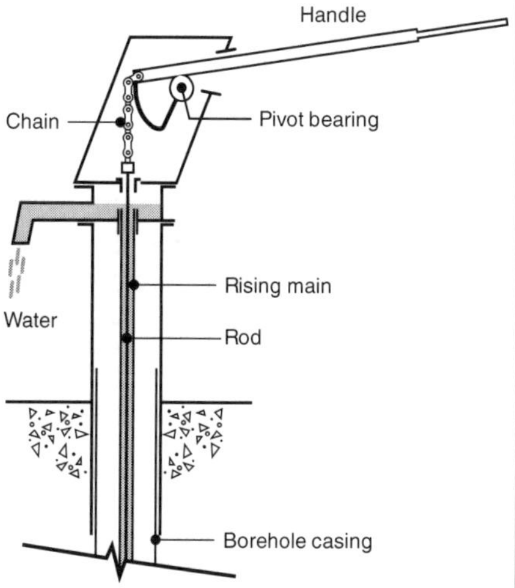
Handle
Chain
Pivot bearing
Rising main
Water
Rod
Borehole casing

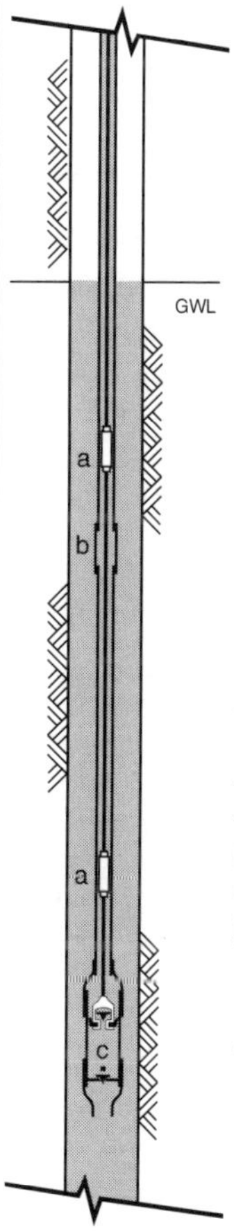
GWL
a
b
a
c

Rising main and cylinder: Traditionally, the rising main is of galvanized steel pipe with a smaller diameter than the piston. The string of pipes and operating rods have to be lifted so that the rod joints (a) and pipe joints (b) can be unscrewed section by section to reach the cylinder (c). This operation needs strong people with appropriate lifting and clamping tools, or a mechanized lifting system. Some manufacturers now supply, therefore, lightweight, thin-walled stainless-steel pipes joined with 'rope threads', or plastic pipes with special threaded collars to reduce the weight which needs to be lifted. Rubber 'O' rings can be used to make such joints watertight.

GWL: Groundwater level

Open-top cylinder design

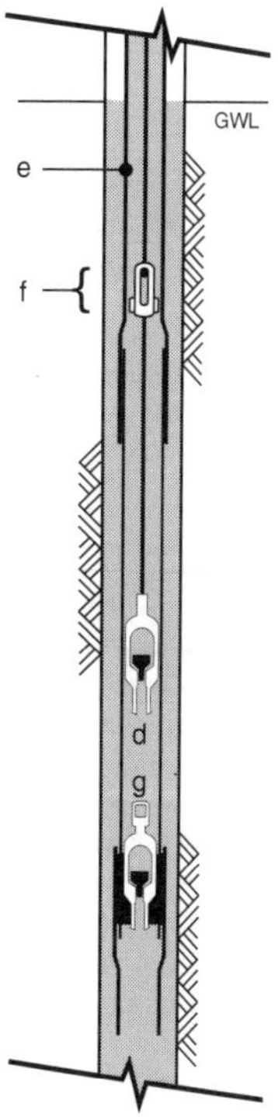
GWL
e
f
d
g

Cylinder: Recent deep-well pump designs have 'open top cylinders' (OTC). These allow the piston (d) to be pulled up through the rising main (e) which is of the same or, preferably, a slightly larger, diameter than that of the cylinder. With these pumps, the piston can be pulled to the surface by pulling out the string of rods.

Rods: Most rod strings are joined by threaded couplings, but some pumps use special rod joints (f) which can be easily disassembled without tools.

GWL: Groundwater level

Foot valve: The best designs of OTC allow the foot valve (g) to be removed through the rising main, either with the piston, or by using a fishing tool which is lowered down inside the rising main on a piece of rope after the piston has been removed.

Rising main removal: In OTC pumps with extractable foot valves, the rising main should never need removing unless the pipe or the lining to the cylinder becomes damaged. Mains with screwed couplings are easily removed.

Should the removal of a solvent-cemented plastic rising main be necessary, the whole length can be removed by supporting it with tall poles so that it can bend to a large radius curve as it leaves the borehole.

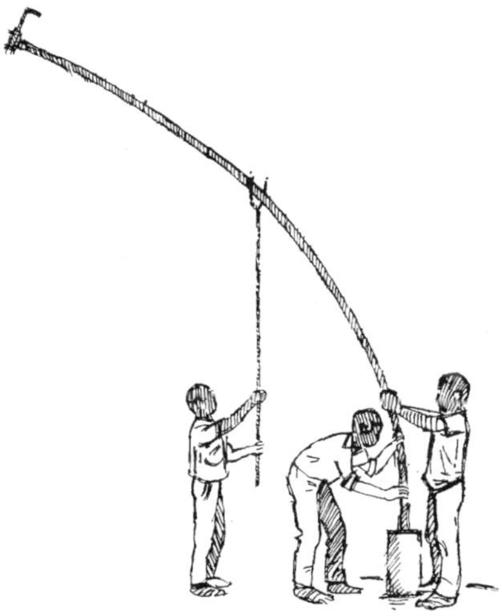

VLOM pumps

Direct action pumps

Main advantages
- Easy access to piston (and sometimes the foot valve), which can be pulled through the rising main.
- Relatively cheap, and easy to manufacture.

Main disadvantages
- Lack of lever handle makes it difficult to operate at pumping lifts much above 12m.
- Pump design is often not rugged enough for use by more than about 50 people per day unless it is frequently repaired.

Deep-well pumps

Traditional design

Main advantages
- Pump is suitable for a wide range of pumping lifts.
- Design can be strong enough to cope with intensive use.

Main disadvantages
- It is difficult to get access to the piston and foot valve.

Open-top cylinder design

Main advantages
- Easy access to piston, and often to the foot valve.
- Use of solvent-cemented plastic rising main is feasible.
- Same advantages as for traditional design.

Main disadvantages
- Large diameter rising main (to allow piston extraction) can be expensive.

Other good features to look for in VLOM pumps:
Corrosion resistance by using:
- stainless steel rods (with deep-well pumps);
- plastic pipe 'rods' (with direct action pumps);
- brass, plastic, and/or rubber for valves and pistons; and
- plastic or stainless steel for the rising main.

Reduction of both production costs and number of different spare parts required by using:
- identical designs for the piston valve and foot valve;
- identical body for piston and foot-valve housing; and
- identical bearings for the rod hanger and handle (can be moulded from engineering plastics).

Few tools necessary for normal maintenance work.
Easily replaceable bearings.
Facility to use 'T' bar end to lever handles to reduce sideways forces on bearings. Handle ideally of adjustable length to suit leverage required.
Theft-resistant parts and 'captive nuts' where possible, so that they cannot be dropped or lost.

Important notes about sustainable maintenance:

Affordability and availability of spares
It is vital that there is a reliable distribution system of essential, affordable spares. Standardizing on one particular pump in a region, or country, can make this, and local technical support for repairs, more feasible.

In-country manufacture
Standardization on one pump in any country can also make the in-country production of a handpump, or at least the spares it commonly requires, a more attractive proposition because of the resulting high level of demand.

Quality control
To give good performance, handpumps and spares need to be produced by manufacturers who carry out stringent quality-control checks.

Vergnet diaphragm pump

This is a deep-well pump which works without rods; instead it uses hydraulic pressure from a small cylinder just under the baseplate of the pump to cause the alternate expansion and contraction of a cylindrical diaphragm in a larger cylinder at the bottom of the borehole. Models for operation by foot or by hand (lever or 'direct action') are available. The reinforced rubber diaphragm can only usually be manufactured in countries with a high level of industrial development.

Special VLOM features:
■ Main wearing parts (in the upper cylinder) are easily accessible.
■ When necessary, the main cylinder can reached by pulling it up using the two flexible plastic pipes attached.

Further reading

Colin, J., *VLOM for Rural Water Supply: Lessons from experience,* WELL, London, 1999. (http://www.lboro.ac.uk/well)
Handpumps: Issues and concepts in rural water supply programmes, IRC Technical Paper No. 25, International Water and Sanitation Centre (IRC), The Hague, 1988.
Arlosoroff S. et al, *Community Water Supply: The Handpump Option*, World Bank, Washington, 1987.
Elson R.J. and Shaw R.J., Technical Brief No. 35: Low-lift irrigation pumps, *Waterlines* Vol.11 No.3, IT Publications, London, 1993.
Franceys R., Technical Brief No. 13: Handpumps, *Waterlines* Vol.6 No.1, IT Publications, London, 1987.
GARNET Handpump Technology Network: http://www.skat.ch/networks/htn/default.htm
Reynolds J., *Handpumps: Toward a Sustainable Technology: Research and development during the Water Supply and Sanitation Decade*, Water and Sanitation Report, UNDP World Bank Water and Sanitation Program, World Bank, New York, 1992.

Prepared by Brian Skinner and Rod Shaw

WEDC Loughborough University Leicestershire LE11 3TU UK
www.lboro.ac.uk/departments/cv/wedc/   wedc@lboro.ac.uk

42. Small-scale irrigation design

Small-scale irrigation can be defined as irrigation, usually on small plots, in which small farmers have the controlling influence, using a level of technology which they can operate and maintain effectively. Small-scale irrigation is, therefore, farmer-managed: farmers must be involved in the design process and, in particular, with decisions about boundaries, the layout of the canals, and the position of outlets and bridges. Although some small-scale irrigation systems serve an individual farm household, most serve a group of farmers, typically comprising between 5 and 50 households.

Small-scale irrigation covers a range of technologies to control water from floods, stream-flow, or pumping:

Flood cropping
- Rising flood cropping (planted before the flood rises).
- Flood/tide defence cropping (with bunds).

Stream diversion (gravity supply)
- Permanent stream diversion and canal supply.
- Storm spate diversion.
- Small reservoirs.

Lift irrigation (pump supply)
- From open water.
- From groundwater.

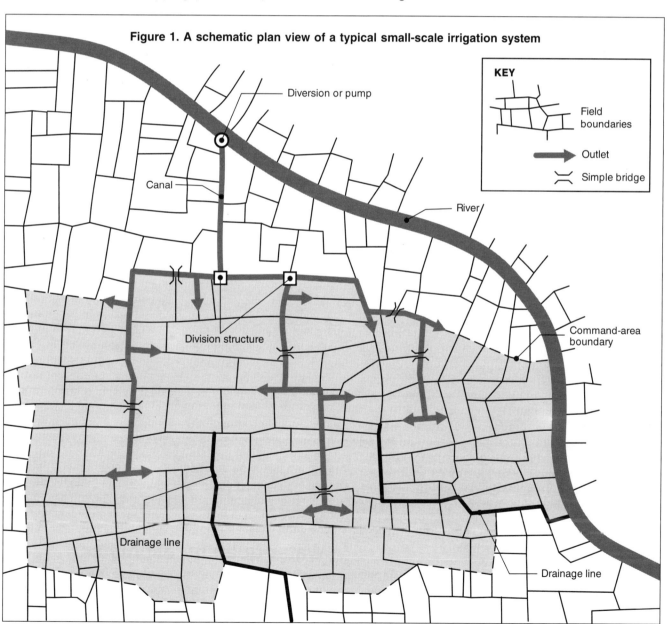

Figure 1. A schematic plan view of a typical small-scale irrigation system

37

Small-scale irrigation design

Water requirements and irrigable area

Crops require a large amount of water for irrigation, and it is important to calculate water requirements accurately, both to design the supply canal and the pump (if any), and to check that enough water is available from the source.

The amount of water required by a crop depends on the local environment, the climate, the crop and its stage of growth, and the degree to which the crop may be stressed. This requirement may be expressed as a uniform depth of water over the area in millimetres per day (mm/d).

Irrigation requirements

Reference evapotranspiration (ET_o) is the water use of grass (in mm/d) under standard conditions. Local estimates may be available from meteorological offices. Typical values are shown in Table 1. For most crops, the reference evapotranspiration at mid-season can be taken as a reasonable estimate of the peak water requirement.

It is reasonable to assume that 70 per cent of average rainfall is available to the crop; the net irrigation requirement (I_n mm/d) can be estimated as:

$$I_n = ET_o - (0.70 \times P)$$

where P (mm/d) is the average rainfall. If a personal computer is available, then the reference evapotranspiration and net irrigation requirements can be estimated conveniently and accurately using the FAO CROPWAT program and the CLIMWAT database.

Additional water has to be supplied to take account of field-application losses which, with surface irrigation, are typically about 40 per cent, giving an application efficiency of 0.60. The field irrigation requirement (I_f) can be estimated as:

$$I_f = \frac{I_n}{0.60} = \frac{ET_o - (0.70 \times P)}{0.60}$$

The field irrigation requirement represents the rate (in mm/d) at which water must be delivered to the field to prevent the crop suffering a shortage of water.

Design command area

The required canal discharge depends on the field area to be irrigated (known as the 'command area'), and the water losses from the canal. For a design command area A (m^2), the design discharge required Q (l/s) for irrigation hours (H) every day, is given by the field-irrigation requirement multiplied by the area, divided by the time (in seconds):

$$Q = \frac{I_f \times A}{H \times 60 \times 60} \quad \text{plus canal losses}$$

Irrigation-canal losses

Water is lost from canals by seepage through the bed and banks of the canal, leakage through holes, cracks and poor structures, and overflowing low sections of bank. The canal losses depend on the type of canal, materials, standard of construction and other factors, but are typically about 3 to 8 litres per second (l/s) per 100 metres for an unlined earthen canal carrying 20 to 60 l/s. Losses often account for a large proportion of water requirements in small-scale irrigation, and may be estimated by 'ponding' water in a trial length of canal, and then measuring the drop-of-water level. When the water-surface width in the canal is W metres, a drop of S millimetres per hour corresponds to an average canal loss of:

$$\frac{W \times S}{60 \times 60} \quad \text{l/s per metre length}$$

Example: What design discharge is required for a canal to irrigate an area of 10 hectares in the semi-arid subtropics, when the mean daily temperature is 30°C, and the mean rainfall is 0.2 mm/d during the peak period (mid-season)? The canal is 800m long and is to operate for 12 hours per day.

Losses from a similar canal are measured as 48mm per hour with a water-surface width of 1.5m.

$$ET_o = 7.5 \text{ mm/d; (see Table 1)}$$

Hence the net irrigation requirement is:

$$7.5 - (0.7 \times 0.2) = 7.36 \text{ mm/d;}$$

and the field irrigation requirement is:

$$7.36/0.60 = 12.3 \text{ mm/d}$$

Canal losses = $\frac{48 \times 1.5}{60 \times 60}$ = 0.02 l/s per metre length

A = 10ha = 10 × 10 000 m^2

Q = $\frac{12.3 \times (10 \times 10\,000)}{12 \times 60 \times 60}$ + 800 × 0.02

= 28 + 16 = 44 l/s

This design discharge of 44 l/s should be compared with the water available from the source. If less is available, the area may need to be reduced, or the irrigation time increased.

Water-quantity estimates

Discharge may be measured using a float, a stopwatch, and tape (for a river), or a weir with a stick gauge (for a small stream or borehole). Technical Brief No. 27 gives details of these methods.

Small-scale irrigation design

Canal design

Water may be conveyed from the source to the field by unlined or lined canal; pipeline; or a combination of the two. The unlined canal is the most common method in use.

A typical cross-section of an unlined earthen canal for small-scale irrigation is shown in Figure 2. To minimize losses, the canal banks should be built from clayey soil and constructed in layers, with each layer compacted using heavy rammers.

The required size of the canal can be decided using Manning's formula:

$$Q = \frac{A \times R^{0.67} \times s^{0.5}}{n}$$

- Q = discharge (m³/s. Note: 1 m³/s = 1000 l/s)
- A = wetted area (m²)
- R = hydraulic radius (m) (= wetted area/wetted perimeter)
- s = slope (fraction)
- n = Manning's roughness coefficient (commonly taken as 0.03 for small irrigation canals)

A design chart, such as Figure 3, can be used.

For example, for a trapezoidal canal in clay soil with side slopes of 1 to 1.5, a design discharge of 44 l/s, and a slope of 0.001 (or 1 m/km), use a bed-width (B) of 0.5 m, and a depth (D) of 0.25 m.

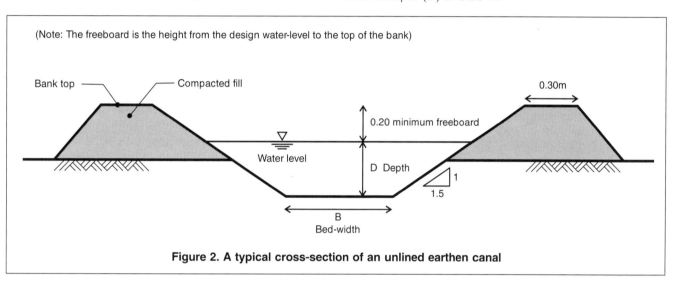

Figure 2. A typical cross-section of an unlined earthen canal

Table 1. Evapotranspiration (ET$_o$) in mm per day for different agro-climatic conditions (FAO, 1977)

Regions	ET$_o$ in mm per day		
Mean daily temperature	<10°C	20°C	>30°C
Tropics			
Humid	3-4	4-5	5-6
Sub-humid	3-5	5-6	7-8
Semi-arid	4-5	6-7	8-9
Arid	4-5	7-8	9-10
Sub-tropics			
Summer			
Humid	3-4	4-5	5-6
Sub-humid	3-5	5-6	6-7
Semi-arid	4-5	6-7	7-8
Arid	4-5	7-8	10-11
Winter			
Humid - sub-humid	2-3	4-5	5-6
Semi-arid	3-4	5-6	7-8
Arid	3-4	6-7	10-11
Temperate			
Humid - sub-humid	2-3	3-4	5-7
Semi-arid - arid	3-4	5-6	8-9

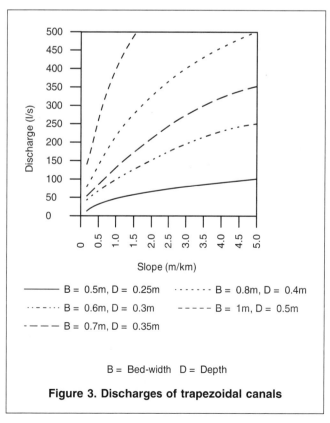

Figure 3. Discharges of trapezoidal canals

Small-scale irrigation design

Distribution outlets

Outlets or division structures are used to distribute the water among a group of farmers. If the flow is less than 30 l/s, one farmer can probably use it efficiently for surface irrigation through one outlet, but larger flows need to be divided fairly between several outlets. In either case, it is important that outlets can be closed when not in use, and that water cannot leak out. The outlet shown in Figure 4 uses a pre-cast concrete, circular gate which has proved to be effective in various countries, lasting longer than either a wood or metal gate. Simple bridges made from planks of wood or a concrete slab are also needed (see Figure 1), so that people and animals can cross the canal without damaging it.

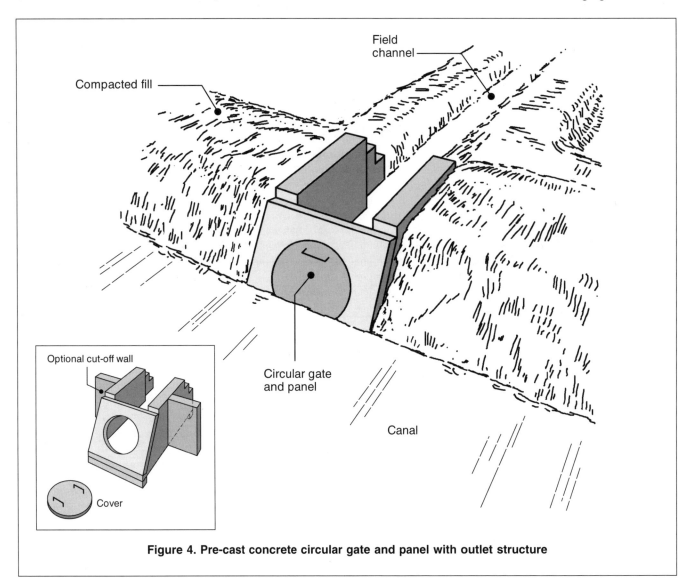

Figure 4. Pre-cast concrete circular gate and panel with outlet structure

Further reading

CLIMWAT for CROPWAT: A climatic database for irrigation planning and management, FAO Irrigation and Drainage Paper No. 49, FAO, Rome, 1993.

Dupriez, H. and de Leener, P., *Ways of Water: Runoff, irrigation and drainage,* CTA/Terres et Vie/Macmillan, 1992.

FAO, Crop Evapotranspiration, FAO Irrigation and Drainage Paper No. 56, Food and Agriculture Organization of the United Nations (FAO), Rome, 1999.

Smout, I.K., Technical Brief No. 27: Discharge measurements and estimates, *Waterlines,* Vol. 9, No.3, IT Publications, London, 1991.

Stern, P., *Small-Scale Irrigation,* IT Publications, London, 1979.

Prepared by Ian Smout and Rod Shaw

WEDC Loughborough University Leicestershire LE11 3TU UK
www.lboro.ac.uk/departments/cv/wedc/   wedc@lboro.ac.uk

43. Simple drilling methods

In many developing countries, water is obtained from handpumps installed above shallow (less than 60m deep) boreholes. It can be expensive to drill the borehole, however, if traditional machine-drilling rigs are used. This Technical Brief outlines simple, low-cost drilling methods which may be used in various situations. Each can be used and maintained easily.

Drilling constraints

Whatever drilling method is used, there are several considerations which must be taken into account:

- The amount of energy required to drill is governed by the rock type. Unconsolidated formations such as sand, silt or clay are weak and much easier to drill than consolidated rocks such as granite, basalt or slate which are hard, strong and dense.
- For hard rocks, cutting tools will need cooling and lubrication.
- Rock cuttings and debris must be removed.
- Unconsolidated formations will require support to prevent the hole from collapse.

Drilling methods

The following low-cost, appropriate drilling methods are described and illustrated on the following pages:

- Percussion drilling
- Hand-auger drilling
- Jetting
- Sludging
- Rotary-percussion drilling
- Rotary drilling with flush

The table below may be used as a guide in the selection of the most appropriate drilling method.

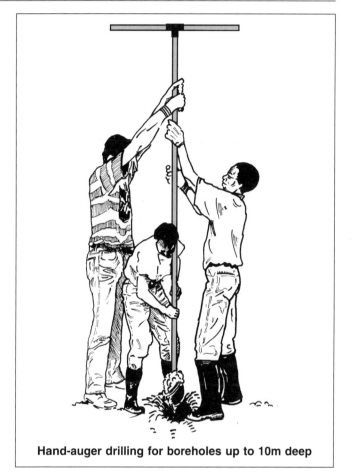

Hand-auger drilling for boreholes up to 10m deep

Table 1. Drilling-method selection		Percussion drilling	Hand-auger drilling	Jetting	Sludging	Rotary percussion drilling	Rotary drilling with flush
Gravel	Unconsolidated formations	✓?	✗	✗	✗	✓?	✗
Sand		✓?	✓	✓	✓	✓?	✓
Silt		✓?	✓	✓	✓	✓?	✓
Clay		✓ slow	✓	?	✓	✓ slow	✓
Sand with pebbles or boulders		✓?	✗	✗	✗	✓?	✗
Shale	Low to medium-strength formations	✓	✗	✗	✗	✓ slow	✓
Sandstone		✓	✗	✗	✗	✓	✓
Limestone	Medium to high-strength formations	✓ slow	✗	✗	✗	✓	✓ slow
Igneous (granite, basalt)		✓ slow	✗	✗	✗	✓	✗
Metamorphic (slate, gneiss)		✓ v slow	✗	✗	✗	✓	✗
Rock with fractures or voids		✓	✗	✗	✗	✓	✓!
Above water-table		✓	✓	?	✗	✓	✓
Below water-table		✓	?	✓	✓	✓	✓

✓ = Suitable drilling method   ✓? = Danger of hole collapsing   ✓! = Flush must be maintained to continue drilling   ? = Possible problems   ✗ = Inappropriate method of drilling

Simple drilling methods

Percussion drilling

Method: The lifting and dropping of a heavy (50kg+) cutting tool will chip and excavate material from a hole. This drilling method has been used in China for over 3000 years. The tool can be fixed to rigid drill-rods, or to a rope or cable. With a mechanical winch, depths of hundreds of metres can be reached.

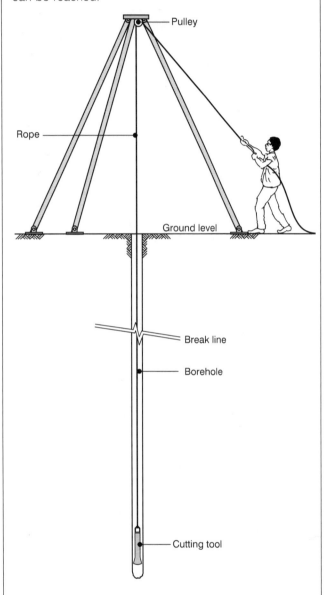

Advantages of percussion drilling:

- Simple to operate and maintain.
- Suitable for a wide variety of rocks.
- Operation is possible above and below the water-table.
- It is possible to drill to considerable depths.

Disadvantages of percussion drilling:

- Slow, compared with other methods.
- Equipment can be heavy.
- Problems can occur with unstable rock formations.
- Water is needed for dry holes to help remove cuttings.

Hand-auger drilling

Method: The cutting tool (known as the auger head) is rotated to cut into the ground, and then withdrawn to remove excavated material. The procedure is repeated until the required depth is reached. Note: This method is only suitable for unconsolidated deposits.

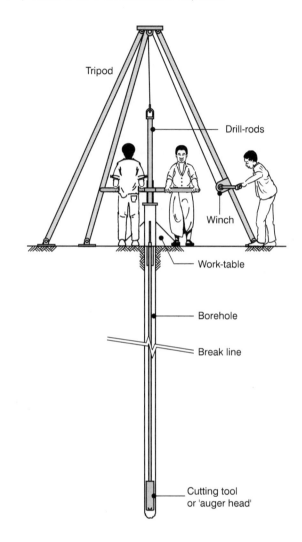

Advantages of hand-auger drilling:

- Inexpensive.
- Simple to operate and maintain.

Disadvantages of hand-auger drilling:

- Slow, compared with other methods.
- Equipment can be heavy.
- Problems can occur with unstable rock formations.
- Water is needed for dry holes.

Useful contacts:

Van Reekum Materials b.v., 115 Kanaal Noord, PO Box 98, AB Apeldoorn, The Netherlands.
Tel: +31 555 335466   Fax: +31 555 313335

V & W Engineering Ltd. (Vonder Rig), PO Box 131, Harare, Zimbabwe. Tel: +263 4 64365/63417
Fax: +263 4 64365

Simple drilling methods

Jetting

Method: Water is pumped down the centre of the drill-rods, emerging as a jet. It then returns up the borehole or drill-pipe bringing with it cuttings and debris. The washing and cutting of the formation is helped by rotation, and by the up-and-down motion of the drill-string. A foot-powered treadle pump or a small internal-combustion pump are equally suitable.

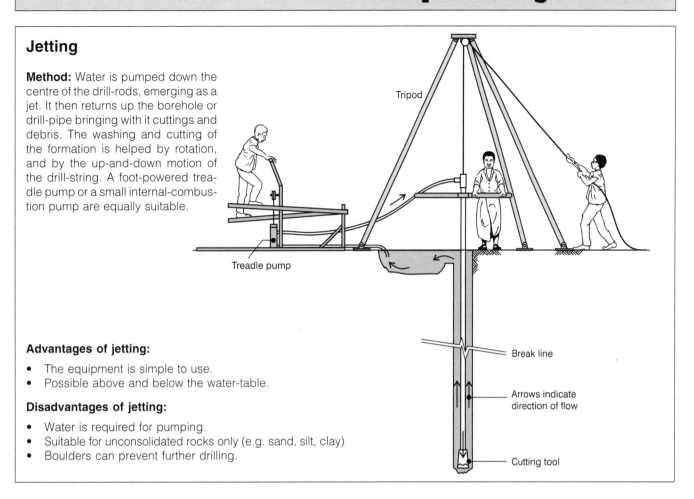

Advantages of jetting:
- The equipment is simple to use.
- Possible above and below the water-table.

Disadvantages of jetting:
- Water is required for pumping.
- Suitable for unconsolidated rocks only (e.g. sand, silt, clay)
- Boulders can prevent further drilling.

Sludging (reverse jetting)

Method: This method has been developed and used extensively in Bangladesh. A hollow pipe of bamboo or steel is moved up and down in the borehole while a one-way valve – your hand can be used to improvise successfully – provides a pumping action. Water flows down the borehole annulus (ring) and back up the drill-pipe, bringing debris with it. A small reservoir is needed at the top of the borehole for recirculation. Simple teeth at the bottom of the drill-pipe, preferably made of metal, help cutting efficiency.

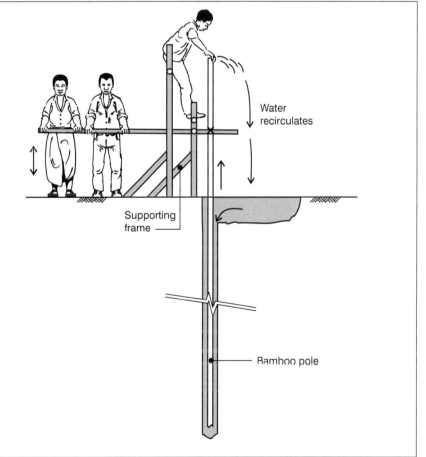

Advantages of sludging:
- The equipment can be made from local, low-cost materials, and is simple to use.

Disadvantages of sludging:
- Water is required for pumping.
- Suitable for unconsolidated rocks only.
- Boulders can prevent further drilling.

Simple drilling methods

Rotary-percussion drilling

Method: In very hard rocks, such as granite, the only way to drill a hole is to pulverize the rock, using a rapid-action pneumatic hammer, often known as a 'down-the-hole hammer' (DTH). Compressed air is needed to drive this tool. The air also flushes the cuttings and dust from the borehole. Rotation of 10-30 rpm ensures that the borehole is straight, and circular in cross-section.

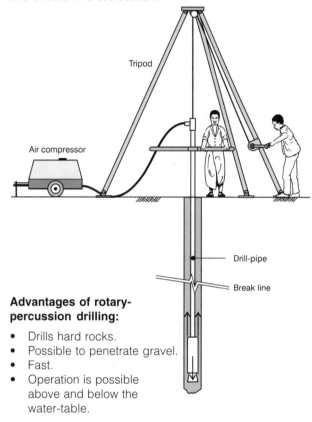

Advantages of rotary-percussion drilling:

- Drills hard rocks.
- Possible to penetrate gravel.
- Fast.
- Operation is possible above and below the water-table.

Disadvantages of rotary-percussion drilling:

- Higher tool cost than other tools illustrated here.
- Air compressor required.
- Requires experience to operate and maintain.

Useful contacts:

Consallen Group Sales Ltd., 23 Oakwood Hill Industrial Estate, Loughton, Essex, IG10 3TZ, UK. Tel/Fax: +44 81 508 5006

Eureka UK Ltd., 11 The Quadrant, Hassocks, West Sussex BN6 8BP, UK. Tel: +44 273 846333 Fax: +44 273 846332

Rotary drilling with flush

Method: A drill-pipe and bit are rotated to cut the rock. Air, water, or drilling mud is pumped down the drill-pipe to flush out the debris. The velocity of the flush in the borehole annulus must be sufficient to lift the cuttings.

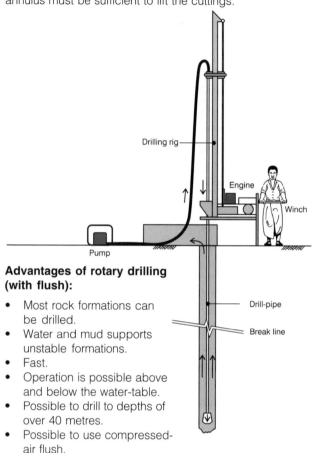

Advantages of rotary drilling (with flush):

- Most rock formations can be drilled.
- Water and mud supports unstable formations.
- Fast.
- Operation is possible above and below the water-table.
- Possible to drill to depths of over 40 metres.
- Possible to use compressed-air flush.

Disadvantages of rotary drilling (with flush):

- Requires capital expenditure in equipment.
- Water is required for pumping.
- There can be problems with boulders.
- Rig requires careful operation and maintenance.

Useful contacts:

Eureka UK Ltd., 11 The Quadrant, Hassocks, West Sussex BN6 8BP, UK. Tel: +44 273 846333 Fax: +44 273 846332

PAT Co. Ltd., 1016 Taskin Road, Thonburi, Bangkok, Thailand. Tel: +66 2 476 1845 Fax: +66 2 476 5316

Further reading

Allen, D.V., *Low-Cost Hand Drilling*, Consallen Group Sales Ltd., Loughton, 1993.
Ball, P., *Bringing Water to the People: Technical brochure*, Eureka UK Ltd., Hassocks, 1994.
Mutwalib, W., *Evaluation of the Muyembe Rural Water Supply*, Loughborough University of Technology, Loughborough, 1994.

Prepared by Bob Elson and Rod Shaw

WEDC Loughborough University Leicestershire LE11 3TU UK
www.lboro.ac.uk/departments/cv/wedc/   wedc@lboro.ac.uk

44. Emergency water supply

This is an overview of the principles of providing water in emergency situations. It outlines the planning procedures necessary for ensuring adequate supply, rather than focusing on design.

As well as food, shelter, and medical aid, providing clean water is usually one of the highest priorities in the event of an emergency. It should be considered alongside immediate sanitation measures, however, which are just as important in controlling many of the most common diseases found in disaster situations (see pages 21-24).

Responding to an emergency

An appropriate response to an emergency depends on whether the emergency affects people where they live, such as in the case of an earthquake or flood, or whether the population is displaced as a result of other pressures such as famine or war. Although the measures may differ, planning considerations for water supply are similar in both situations.

Planning what to do

There are several planning considerations:

- **Demand assessment**
  How much water is needed?
- **Location and protection of water sources**
  Where are the nearest/most convenient sources of water? How can they be protected?
- **Water treatment**
  What is the level of water treatment required for use/consumption? What methods are available for treatment?
- **Water distribution**
  Where will people collect water from?
- **Collection and storage**
  How will the water be collected? How will it be stored for domestic use?

These questions must be considered as soon as an emergency occurs. It is also important to prioritize action. A phased action plan comprises:

- Immediate measures (to sustain life);
- intermediate measures (from about 2 to 6 weeks after the disaster, or the arrival of refugees in a camp); and
- long-term solutions (from about 6 weeks).

The actual duration of these phases is usually determined by three factors:

- Accessibility of the disaster area or refugee camp for local, national, and international assistance;
- the nature of the disaster; and
- the availability of water, materials, and skilled labour.

A refugee camp

Emergency water supply

Demand assessment

Demand estimates will clearly depend on local conditions. Table 1 shows guideline figures only. It is important to note that water quantity alone is not sufficient to ensure the health of refugees. Good sanitation and hygiene education and behaviour are also essential.

Table 1. A guide for assessing the demand for water in a disaster situation

Individuals	Minimum for survival	3 – 5 l/p/d
	Desirable emergency supply	15 – 20 l/p/d
Health centres	Out-patients only	5 l/patient/d
	In-patients (excluding cholera hospitals) (not including laundry)	40 – 60 l/patient/d
Feeding centres		20 – 30 l/p/d
Toilet flushing water	Pour flush latrines	2 – 8 l/p/d
	Cistern flush	40 – 50 l/p/d
Animals (approx.)	Cattle	20 – 30 l/h/d
	Horses, mules, donkeys	15 – 25 l/h/d
	Sheep, goats,	10 – 20 l/h/d
	Camels	2 l/h/d
Irrigation	Very variable, but typically	3 – 6 l/m^2/d

Allow at least 40% extra for unforeseen circumstances and waste.

Location of water sources

There are three types of water source:

- **Existing sources**
  When disasters occur where people live, it may be possible to revive some or all of the existing water supplies.

Existing water sources may be revived

- **Local sources**
  Local sources of water in areas where existing supplies are inaccessible due to the nature of the disaster, or where refugees gather away from existing communities; may include wells, boreholes, springs, streams, ponds or rainwater.

- **Distant sources**
  Water will usually be available from existing communities — which may be some distance away from the disaster area or refugee camp. The impact on these communities of an increase in the volume taken from a water source must always be considered.

A note about water quality

Normal practice is to supply all water of a quality suitable for drinking.

Treating water to this standard takes time — even using the simplest of methods. It is often appropriate, therefore, to supply treated water from a distant source as an immediate measure, where possible. At this stage, cost is not usually a problem, as both government and non-governmental organizations are generally willing to mobilize short-term resources promptly.

The most common form of supplying water in this way is by water tanker. Though expensive, tankers can be mobilized quickly and offer flexibility of distribution.

Nevertheless, immediate consideration should be given to alternative sources of water and ways of treating this water once it is located.

Water tankers

Emergency water supply

Water treatment (surface sources)

At the outset, it is usually unrealistic to expect the quality of water supplied to satisfy normal water-quality guidelines. Provided that the water is clear to look at, does not smell or have an unpleasant taste, and is disinfected, it is usually acceptable in the short-term, but it should be tested as soon as possible. There are a number of suitable water-treatment options to consider as longer-term solutions. These should aim to provide better-quality water.

■ Infiltration wells

The sand and gravel deposited beside and below a river acts as a very effective water filter. So, wells dug a short distance from a river-bank will usually provide better-quality water than the river itself.

Digging an infiltration well

■ Settlement

The quality of water from surface-water sources, such as streams and rivers, can be increased significantly by allowing water to stand in calm conditions, preferably under cover. This allows some of the suspended material and associated pollution to sink to the bottom. Some fine particles will not settle, but will remain suspended indefinitely, unless encouraged to settle by the addition of small quantities of certain chemicals, such as aluminium sulphate. This must take place under carefully controlled conditions, as too much chemical additive will poison the water, and too little will not achieve the desired result of purifying the water.

Allowing water to settle in a package storage tank

■ Filtration

For large populations, slow-sand filters provide one of the simplest and most reliable forms of water treatment, but they occupy large areas of land, and require careful design and maintenance.

Small volumes of drinking-water, suitable for individual households, can be obtained from domestic filters that allow water to pass through ceramic filter 'candles'.

Domestic water filters

■ Disinfection

As a final precaution to ensure that water is bacteriologically safe, it should be disinfected. This reduces the number of bacteria present in the water to a safe level. Disinfection is most effective in clear water. Chlorination is the most widely used technique, as it is available in various forms and as some of the chlorine compound should remain in the water, increasing the likelihood that the water will remain safe to drink during distribution and storage.

■ Package water-treatment plants

Package water-treatment plants are highly mechanized, self-contained units which, though small, compact, and quick to install, are expensive, and require routine maintenance by a skilled operative. They have been used successfully in Turkey and Zaire, however, as one of several water-treatment options.

A package water-treatment plant

Emergency water supply

Water distribution

As noted above, water tankers are usually only suitable as a short-term measure. Water may be distributed more efficiently using a simple pipe network. Where possible, the distribution network should be connected to a water-storage tank. The tank may be filled slowly over a 24-hour period — allowing people to draw water when they need it most. The pipes should be designed to carry the maximum expected flow (see *The Worth of Water*, pages 117-120).

The water-collection point usually comprises a series of taps attached to the end of the pipe network (see right). Tap stands should be evenly distributed throughout the camp, and must be strong enough to withstand heavy, sometimes continuous, use.

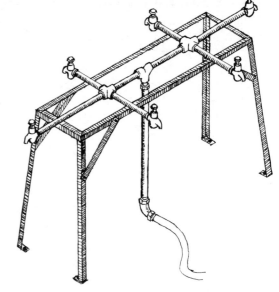

Oxfam water-collection point

Collection and storage

The collection and storage of water by refugees for domestic use is an important consideration both in the immediate aftermath of a disaster, and in the long-term. Refugees may have no means of collecting water from its source, so containers must be provided.

Contamination of water often takes place *after* the water has been collected from the supply. It is necessary, therefore, to provide water-collection and storage vessels that are easy to use and keep clean.

Hygiene education has an important role to play in ensuring that water is not contaminated.

Covered water containers

Further reading

Davis, J. and Lambert, R., *Engineering in Emergencies: A practical guide for relief workers*, IT Publications, London, 1995.
House, S. and Reed, R.A., *Emergency Water Sources: Guidelines for selection and treatment*, WEDC, Loughborough, 1997.
Médecins Sans Frontières, *Public Health Engineering in Emergency Situations*, Paris, 1994.
Oxfam Water Supply Scheme for Emergencies, Oxfam Technical Unit, Oxford, 1992.
Reed, R.A., 'Technical Brief No. 29: Designing simple pipelines', *Waterlines*, Vol.10, No. 1, IT Publications, London, 1991.
Reed, R.A. and Shaw, R.J., 'Technical Brief No. 38: Emergency sanitation for refugees', *Waterlines*, Vol.12, No. 2, IT Publications, London, 1993.
Reed, R.A., (ed.) *Technical Support for Refugees: Lessons from recent experiences*, Proceedings of the 1991 International Conference, WEDC, Loughborough University of Technology, Loughborough, 1993.
Reed., R.A. and Smith, M.D., 'Water and sanitation for disasters', *Tropical Doctor*, Vol. 21, Supp. No. 1, Royal Society of Medicine, London, 1991.
UNHCR, *Handbook for Emergencies*, Geneva, 1982.
UNHCR, *Water Manual for Refugee Situations*, Geneva, 1992.
Waterlines, Vol.13, No.1, IT Publications, London, 1994.

Prepared by Bob Reed and Rod Shaw

WEDC Loughborough University Leicestershire LE11 3TU UK
www.lboro.ac.uk/departments/cv/wedc/   wedc@lboro.ac.uk

45. Latrine slabs and seats

The most common type of sanitation for low-income communities, by far, is a pit latrine. This Technical Brief introduces some of the features of latrine slabs and seats which help to improve the safety and comfort of users.

The pit latrine

Excreta decomposes in the pit, forming:

- **gases** (which escape to the atmosphere, or are absorbed in the soil around the pit);
- **liquids** (which seep into the soil beneath and around the pit); and
- **a solid residue** (which accumulates).

Figure 1 shows a section of a simple pit latrine with a latrine slab. Many people, however, (especially in Asia) wash themselves with water after defecation. They usually defecate into a pan which is then cleaned by pouring a litre or so of water from a bucket (*lotta*). This type of pan is often known as a 'pour-flush' pan. A water seal (or trap) prevents smells, and stops insects from leaving the pit (Figure 2a).

A pan above a pit is usually supported by a slab, and the trap hangs down above the pit (unless offset). This trap is sometimes called a 'gooseneck'.

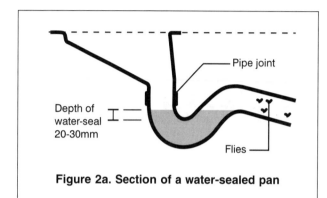

Figure 2a. Section of a water-sealed pan

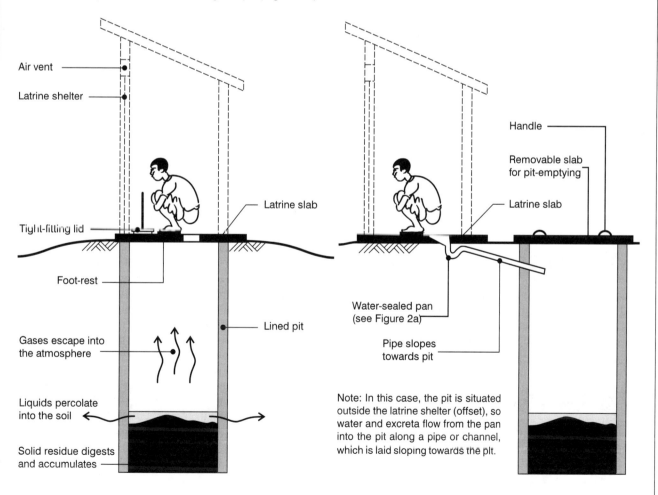

Figure 1. Section of a simple pit latrine

Figure 2. Section of an offset pour-flush pit latrine

Note: In this case, the pit is situated outside the latrine shelter (offset), so water and excreta flow from the pan into the pit along a pipe or channel, which is laid sloping towards the pit.

Latrine slabs and seats

Using the latrine

If people squat to relieve themselves, they usually put their feet on a **slab** which goes across the top of the pit. Faeces and urine fall through a squat-hole which is constructed so that the user's feet rest either side.

Other people, either because of custom, or simply because they find it more comfortable, prefer to sit on a **seat** when using a latrine.

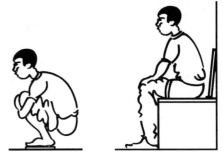

Figure 3. Using a slab and using a seat

Latrine slabs

An effective way of ensuring that a slab is strong enough to support the heaviest users is for five men to stand on it! (Figure 4).

Latrine seats

A latrine seat is made by building or fixing a support or pedestal on top of the slab. The height depends on what the majority of users find comfortable. This is normally about 350mm above the floor, with the hole commonly 250mm wide.

A seat support can be made with timber, brick, concrete, or blocks. The seat itself is usually made of timber or concrete. A hinged cover may be fitted to control the escape of flies and smells.

Figure 4. Ensuring that a slab is strong

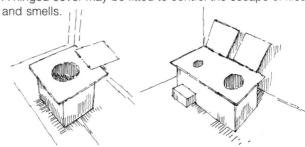

Figure 5. Latrine seats

Seats for children

Children may be frightened by a large opening, so a separate hole with a smaller diameter may be provided.

The seat height may be a problem, so a lower children's seat may be built. Alternatively, a block can be placed for children's feet, as shown above (right).

Slab supports

Latrine slabs should be properly supported. If the slab rests on bare earth (where the pit is not lined), it needs about 200mm-wide support all round. So a 1100mm-diameter pit requires a 1500mm-diameter slab. On a good base — such as the top of a brick lining, 50mm width of support is enough.

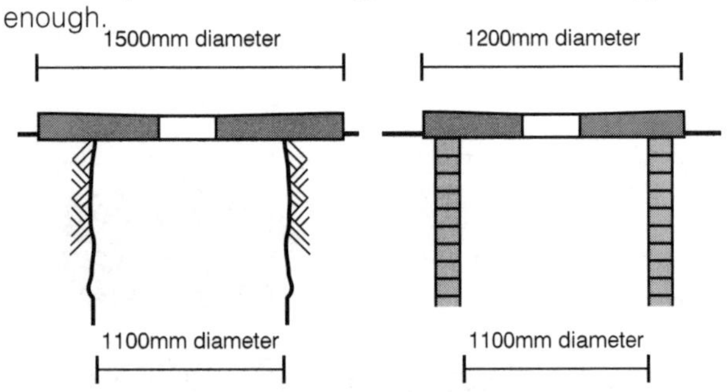

A concrete collar to support the slab may be formed by digging a trench round the top of the pit.

The top of a brickwork or masonry pit lining may be corbelled so that the slab can be reduced in size.

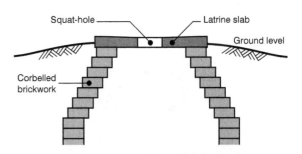

Figure 6. Supporting latrine slabs

Figure 7. Corbelled brickwork

Latrine slabs and seats

More about latrine slabs ...

Squat-holes
A squat-hole is usually about 400mm long; any smaller and fouling is likely. Children must be protected, so the hole should not be too wide. A child is unlikely to fall through a hole that is less than 180mm wide.

Traditionally, squat-holes come in many shapes. The best shape is a 'keyhole' (see Figure 8).

Footrests
Footrests are often provided for both squat-holes and pour-flush pans. They are usually about 10mm above the remainder of the surface of the slab, and keep users' feet off the floor. They also enable users to find a good position for squatting, which is especially helpful at night.

Slab surface
In order to maintain cleanliness, the surfaces of all slabs should be as hard and smooth as possible. Except for domed slabs, the top of the slab should slope towards the hole allowing spilled water, or water used for cleaning, to drain into the pit.

Domed slabs
Slabs can be cast as flat domes with no reinforcement. They are strong enough to support their own weight and the weight of latrine users.

The dome shape is made by mounding earth to the required profile of the underside of the slab. The earth is compacted and smoothed, and then covered with plastic sheeting or old cement bags, or coated in old engine oil.

Domed slabs 1500mm in diameter can be made 40mm thick at the edges, with the centre of the bottom rising 100mm.

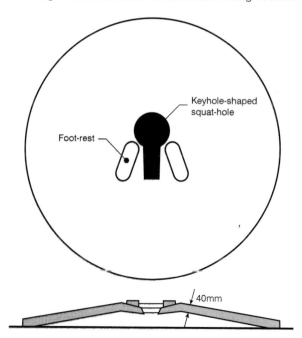

Figure 8. Plan view and cross-section of a domed slab

Small concrete slabs
Squat-holes set in small concrete slabs about 600mm square can make a pole-and-mud slab hygienic.

These slabs, sometimes referred to as SanPlats, weigh about 35kgs, light enough to be carried by one person, and lifted by two.

Figure 9. Positioning a SanPlat

A tight-fitting lid is often used with this type of small slab — an effective way of dealing with flies, smells, and the risk of hookworm transmission. The handle can be wooden or made from a length of reinforcing steel.

A tight-fitting concrete lid can be cast in the squat-hole. Holes may differ slightly, so the lid and its corresponding slab may be given an individual identification mark or number.

Figure 10. A tight-fitting lid

Removable slabs
Of course, the pit has to be emptied, so the whole, or part, of the slab should be easy to lift.

Removable slabs usually come with handles made of reinforcing steel (see Figure 2).

Movable slabs should be sealed with weak mortar (made with mud or lime) to ensure an insect-proof joint.

Large slabs should be constructed in sections, so that each part can be lifted by two men. Each section should be at least 600mm wide, however, to allow enough room to empty the pit.

Latrine slabs and seats

Materials for latrine slabs
Latrine slabs are made from a variety of materials, depending on what is available.

Concrete
Concrete is ideal: it is strong, durable, and has a smooth, easy-to-clean surface.

Slabs can be made *in situ:* a mixture of cement, sand, stones and water is laid over the pit and left to 'cure'. A temporary timber platform to support the concrete is essential.

More often, concrete slabs are **prefabricated** or **pre-cast** — made and cured until strong enough to be moved without cracking or breaking.

Most pre-cast slabs are flat, and are reinforced with steel bars. Mild-steel bars, 6mm in diameter, spaced 150mm apart; or 8mm in diameter and spaced 250mm apart, in both directions, are normally sufficient for an 80mm-thick slab spanning up to 1.5 metres.

The shape of the slab depends on the individual pit: it can be rectangular, square, or circular.

Circular slabs have the advantage that they can be rolled into the correct position.

Ferrocement slabs (slabs made of cement reinforced with wire mesh) may be thinner, lighter, and easier to handle than normal reinforced-concrete slabs.

Two, three, or four layers of chicken wire are plastered with several layers of a rich cement mortar to make a slab about 20mm thick. The mortar is made by mixing one part of cement to two parts sand, and adding enough water to obtain a thick, creamy consistency.

Logs and bamboo
Logs and bamboo are often used as they are locally available and cheap. Sometimes they are free.

Ideally, the harder the wood, the better. Timber should also be protected against rot and termite attack by being soaked in used oil, for example.

Logs and bamboo are often covered with a layer of gravel and/or mud finished to a smooth, hard surface. Traditional methods include using animal dung or cassava (soaked in water overnight) which is then mixed with soil.

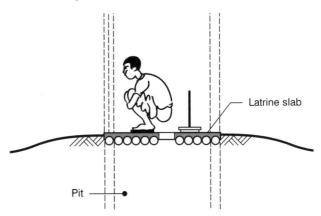

Figure 12. Section of a bamboo slab

Other materials
Other suitable materials are sawn timber, metal sheets (for example, old advertisement signs), natural stone, and brick arches. The chassis of an abandoned car or lorry makes a good support for a slab, and the slab itself may be made from flattened car doors and other scrap material.

Figure 11. Rolling a circular slab

Further reading
Brandberg, B., *Latrine Building: A handbook to implementing the SanPlat system,* IT Publications, London, 1997.
Franceys, R.W.A., Pickford, J.A., and Reed, R.A., *A Guide to the Development of On-Site Sanitation,* World Health Organization, Geneva, 1992.
Morgan, Peter, *Rural Water Supplies and Sanitation,* Macmillan, London, 1991.
Pickford, John, *Low-Cost Sanitation: A survey of practical experience,* IT Publications, London, 1995.
Pickford, John (ed.), *The Worth of Water: Technical Briefs on health, water and sanitation,* IT Publications, London, 1991.

Prepared by John Pickford and Rod Shaw

WEDC Loughborough University Leicestershire LE11 3TU UK
www.lboro.ac.uk/departments/cv/wedc/   wedc@lboro.ac.uk

46. Chlorination

Some water sources contain disease-causing organisms which need to be removed or killed before the water is safe to drink. If carefully undertaken and monitored, disinfection is an effective means of removing such organisms. Chlorine is the most widely used disinfectant, and one which is often the most readily available. This technical brief describes a method of calculating the dose of chlorine required to disinfect small community water supplies.

Why disinfect?

Water treatment processes such as storage, sedimentation, and sand filtration will reduce the content of disease-causing organisms in water, but will not leave it completely free of such organisms. Disinfection, when applied and controlled properly, is the most practical and effective means of removing such organisms.

Methods of disinfection

Boiling water may be effective as a method of disinfection, but it is not practicable for large quantities. Sunlight can also act as a natural method of disinfection, but it is difficult to control and manage. For these reasons, chemical disinfectants (especially chlorine and chlorine compounds) are used. Iodine may also be used as a disinfectant, but it is usually more expensive than chlorine compounds.

Chlorine compounds will destroy disease-causing organisms quickly — usually after 30 minutes. They are widely used and are relatively inexpensive. If carefully applied, chlorine has the advantage that a measurable residual of chlorine in solution can be maintained in the water supply. This residual provides further potential for disinfection and is also an important indicator of successful application.

When to use chlorine

Chlorine may be used:

- when suitable compounds are available and their application can be strictly controlled;
- when there is enough time between the addition of chlorine to water and the consumption of the water;
- where a community has a continuous supply of water, with storage capacity; and
- by individuals to provide additional protection.

When not to use chlorine

Chlorine should not be used:

- when a regular supply of chlorine compounds cannot be guaranteed;
- where chlorine may react with other chemicals in the water creating undesirable or dangerous by-products;
- to attempt to kill cysts or viruses; or
- when careful monitoring cannot be provided.

Chlorine demand and residual

When chlorine is added to a water source, it purifies the water by damaging the cell structure of bacterial pollutants, thereby destroying them. The amount of chlorine needed to do this is called the **chlorine demand** of the water. The chlorine demand varies with the amount of impurities in the water. It is important to realise that the chlorine demand of a water source will vary as the quality of the water varies.

The aim of chlorination is to satisfy the chlorine demand of the water source. Once the demand has been satisfied, any excess chlorine above the level needed to satisfy the demand remains as a **residual** of chlorine (chlorine residual) in the supply.

If a supply is to be adequately disinfected, therefore, there should be a chlorine residual in the supply, so that there is the capacity to cope with any subsequent bacterial contamination. The chlorine residual should generally be in the range 0.3 to 0.5mg of chlorine per litre of treated water. Any more than this and the supply may taste bad and be harmful, and people may refuse to use it. Any less, and there is no guarantee that the supply is adequately protected. An example is given in Figure 1 below.

A water supply (from a spring, for example) has a chlorine demand of 2.0 mg/l.

If exactly 2.0 mg of chlorine is added per litre of water, then the chlorine demand will *just* be met and there will be no chlorine residual.

If 2.5 mg/l of chlorine is added, then the chlorine demand will be met and exceeded, so that a residual of 0.5 mg/l will be left in the water when it goes into supply.

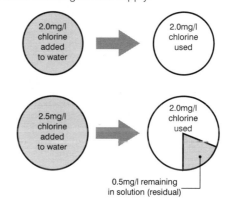

Figure 1. An example of chlorine residual

53

Chlorination

Testing for chlorine residual

The most common test is the dpd (diethyl paraphenylene diamine) indicator test, using a comparator. This test is the quickest and simplest method for testing chlorine residual.

With this test, a tablet reagent is added to a sample of water, colouring it red. The strength of colour is measured against standard colours on a chart to determine the chlorine concentration. The stronger the colour, the higher the concentration of chlorine in the water

Several kits for analysing the chlorine residual in water, such as the one illustrated in Figure 2, are available commercially. The kits are small and portable.

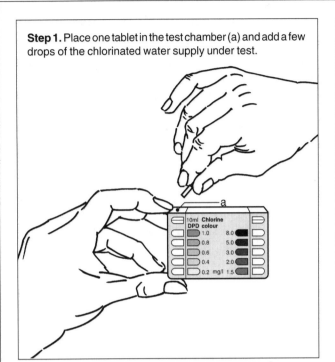

Step 1. Place one tablet in the test chamber (a) and add a few drops of the chlorinated water supply under test.

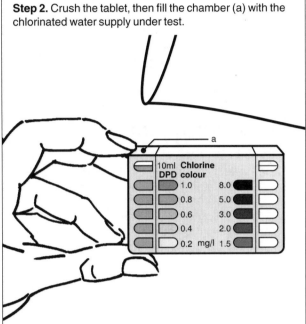

Step 2. Crush the tablet, then fill the chamber (a) with the chlorinated water supply under test.

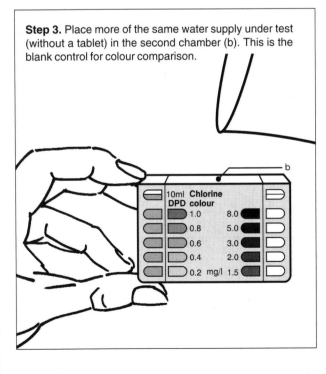

Step 3. Place more of the same water supply under test (without a tablet) in the second chamber (b). This is the blank control for colour comparison.

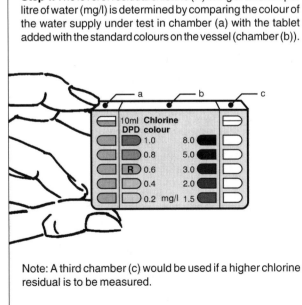

Step 4. The level of residual chlorine (R) in mg of chlorine per litre of water (mg/l) is determined by comparing the colour of the water supply under test in chamber (a) with the tablet added with the standard colours on the vessel (chamber (b)).

Note: A third chamber (c) would be used if a higher chlorine residual is to be measured.

Figure 2. Procedure for testing for chlorine residual

Chlorination

Chlorinating water supplies

Chlorine is available in many forms — as chlorine gas and in compounds such as bleaching powder, high test hypochlorite (HTH), tablets, granules, and liquid bleach.

Each product contains a different amount of usable chlorine, so different quantities of each will be required for the same purpose. In addition, the chlorine content of each product will reduce over time as the source is exposed to the atmosphere. All products should be carefully stored to minimize deterioration.

The best practical method of chlorinating a supply of water is to use two storage tanks of suitable size alternately, one filled from the source, while the other is used for supply.

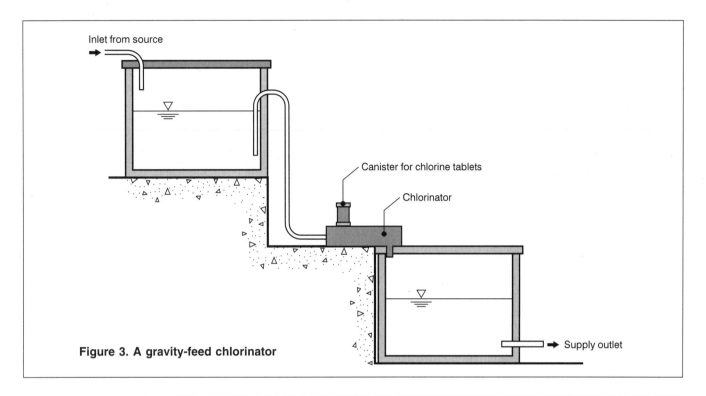

Figure 3. A gravity-feed chlorinator

A chlorination checklist

- Chlorine needs *at least* half an hour contact time with water to disinfect it. The best time to apply it is after any other treatment process, and before storage and use.

- Never apply chlorine before slow sand filtration or any other biological process, as the chlorine will kill off the bacteria which assist treatment, making the treatment ineffective.

- Never add any solid form of chlorine directly to a water supply, as it will not mix and dissolve. Always make up as a paste first, mixing the chlorine compound with a little water.

- Disinfection is only one defence against disease. Every effort should be made to protect water sources from contamination, and to prevent subsequent contamination during collection and storage.

- The correct procedure for applying a disinfectant to water should be strictly adhered to, and water supplies should be monitored regularly to ensure that they are free from bacteria. Otherwise, people may be misled to believe that the water is safe to drink when, in fact, it is hazardous to do so.

- The optimum chlorine residual in a small, communal water supply is in the range of 0.3 to 0.5mg/l.

- The chlorine dose required to disinfect a supply will increase if the water is very turbid. In such circumstances, it is best to treat the water to reduce turbidity *before* chlorination.

Caution

All forms of chlorine are harmful to health — avoid skin contact and do not inhale the fumes. Chlorine should be stored in cool, dark, dry and sealed containers, and *out of reach of children*.

Chlorination

Modified Horrocks' method of chlorination

With most chlorination methods the operator should make up a solution of known concentration. For the reasons outlined on page 55, however, it is not usually possible to do this accurately. The modified Horrocks' method of chlorination can be carried out without prior knowledge of the chlorine content of the chlorine product.

Equipment

- 5 containers (any type — as long as they are all the same size. Plastic drinks bottles may be used).
- A measuring device to measure out the solid chlorine product. An oral rehydration therapy (ORT) spoon would be suitable. The exact size of the spoon is not critical, but identical amounts must be measured out each time, so use a level spoonful, for example. Use a spoon which measures out about 1g, as this will help in the calculations.
- A device to dispense small quantities of liquid (a small 1ml or 5ml syringe would be suitable).
- Dpd test equipment

Method

1. Label the five containers 1 to 5.
2. Place one level spoonful of chlorine product (bleaching powder or HTH) into the first container. If the spoon has a capacity of 1g, there is now 1g of chlorine product in the container.
3. Add a few drops of the water to be chlorinated, and mix to a paste (dissolving the chlorine-product powder).
4. Dilute the paste with enough water to fill the container. If the container holds one litre, it now contains one spoonful per litre, or in this case, one gram per litre. **If we take 1ml out of this container, this will contain 1mg of chlorine product** (1 litre = 1000ml, and 1g = 1000mg).
5. Fill containers 2, 3, 4, and 5 to capacity (1 litre) with the water to be chlorinated.
6. Transfer 2ml of the liquid from container 1 to container 2, 4ml to container 3; 6ml to container 4; and 8ml to container 5. Container 2 will then have 2mg/l; container 3, 4mg/l; container 4, 6mg/l; and container 5, 8mg/l.
7. Leave containers 2, 3, 4 and 5 to stand for at least 30 minutes — this is the minimum *contact time* required for the chlorine to disinfect the water.
8. Test the water in each container for residual chlorine content using the Dpd test kit (see page 54).
9. The container with the lowest concentration of chlorine equal to or more than 0.4mg/l indicates how much chlorine powder should be added to the water being disinfected.

Example

A water supply from a spring with a daily flow of 70 m^3/day needs chlorinating to make it safe to drink. Tests on the water — using the modified Horrocks' method — indicated residual chlorine concentrations (after 30 minutes) of 0, 0.2, 0.5 and 1.0mg/litre in containers 2, 3, 4, and 5 respectively.

Therefore, container 4, with a residual concentration of 0.5mg/l, had the lowest residual chlorine concentration equal to or exceeding 0.4mg/l.

The concentration of chlorine product added to container 4 was 6mg/l. For a supply of 70m^3/day, therefore, the amount of chlorine product to be used is calculated as:

70 x 1000 litres x 6mg/l = 420 000mg = 420g = 0.42kg.

(1m^3 = 1000 litres)

Further reading

Cairncross, S. and Feachem, R.G., (1993). *Environmental Health Engineering in the Tropics: An introductory text,* John Wiley, Chichester.

IRC (1986). *Small Community Water Supplies: Technology of small water supply systems in developing countries.* Technical Paper No. 18, IRC, The Hague.

IRC (1982). *Practical Solutions in Drinking-Water Supply and Waste Disposal for Developing Countries.* Technical Paper No. 20, IRC, The Hague.

Mann, H.T. and Williamson, D., (1993). *Water Treatment and Sanitation,* IT Publications, London.

Twort, A.C., Law, F.M., Crowley, F.W. and Ratnayaka, D.D., (1994). *Water Supply,* Edward Arnold, London.

White, G.C. (1972). *Handbook of Chlorination,* Van Nostran Reinhold, New York.

Prepared by Jeremy Parr, Michael Smith and Rod Shaw

WEDC Loughborough University Leicestershire LE11 3TU UK
www.lboro.ac.uk/departments/cv/wedc/   wedc@lboro.ac.uk

47. Improving pond water

Wherever possible, a community should avoid the health risks which result from using contaminated pond water, by using an alternative, good-quality source. (Groundwater or rainwater sources will usually produce water of much better quality.) If a pond is the *only* source of water, the implementation of some of the ideas in this Technical Brief should improve the quality of the water. It is not easy to find a sustainable way to produce good-quality potable water from ponds, but any improvement in the quality of pond water is worth the effort. This Brief concentrates on the removal of suspended matter and pathogenic organisms. It does not address the removal of chemical contaminants because of the complexities involved.

Health risks

The consumption of untreated pond water is a health risk. Like all surface water sources, it is likely to contain one or more of the following contaminants:

- **Pathogens** (disease-causing organisms, many of which come from faeces) such as bacteria, viruses, protozoa, and guinea-worm larvae. If *schistosome cercaria* are present, they can penetrate the skin of anyone entering the water-causing bilharzia.

- **Harmful chemicals**
  Agricultural — pesticides, herbicides, fertilizers
  Industrial — heavy metals (e.g. chromium)

- **Other contaminants**
  These affect the appearance and taste of the water and make filtration difficult:

  Algae may release poisonous toxins when they die. Their growth is promoted by the presence of fertilizer or other sources of phosphates or nitrates in the water.

  Suspended solids — fine particles of soil, particularly clays, to which bacteria and viruses often become attached. If these particles can be removed (e.g. by settlement), so can many of the pathogens.

Reducing contamination

Contaminants come from many sources (Figure 1). It is easier to prevent them entering the pond than to remove them from the water. To be successful, the following methods of reducing contamination need wide support from the community, plus participatory health education:

- **Restricting activities around the pond**
Whenever possible, fence off the catchment area draining into the pond and prevent polluting activities from taking place within the catchment area.

- **Restricting activities in the pond**
Faecal pathogens enter the water:
- when faeces are deposited in the water;
- when people wash themselves or their clothes; and
- on the feet of people and animals.

Keeping people and animals out of the pond will improve the water quality. It will also prevent the spread of guinea-worm disease.

Some communities may be able to devote one pond to bathing and watering livestock, and leave another pond protected from these activities so that the water quality is maintained.

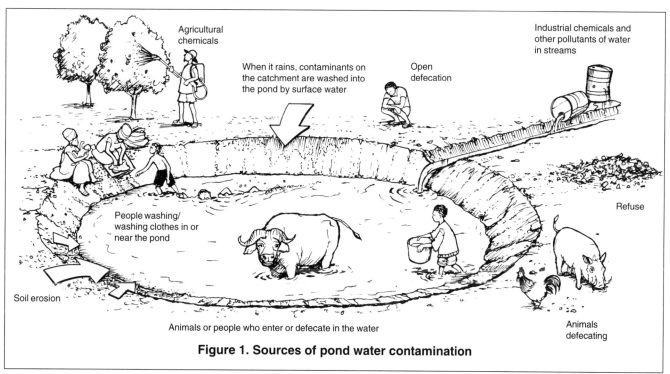

Figure 1. Sources of pond water contamination

Improving pond water

Collecting water without entering the pond

People can only be discouraged from entering the pond if alternative methods for collecting water are provided:

Platforms, steps or ramps (Figure 2) can be used to bring people close enough to the water for them to bend down and fill a bucket, but contamination deposited on these structures can enter the pond, especially when it rains. If the pond level varies considerably, platforms will need to have floating sections to keep the access close to the water. Alternatively, people can draw water by bucket and rope. **Bank-mounted devices** (Figure 3) which keep people well away from the water are ideal, but if a handpump is to be used it must be sustainable. Spilt water should not be allowed to flow back directly into the pond, and is best disposed of into a soakaway.

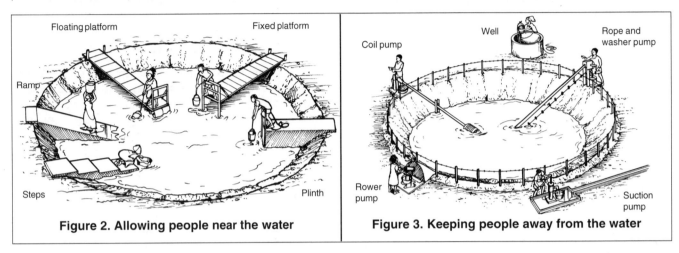

Figure 2. Allowing people near the water

Figure 3. Keeping people away from the water

Drawing strained or filtered water

The water near the surface of a pond, and away from its edges, usually contains less suspended solids than anywhere else in the pond. There are some advantages, therefore, in using a **floating intake pipe** (Method 1). A coarse strainer (such as a **perforated pipe**) will keep out amphibians and plants. A floating **strainer box** has a larger surface area which may permit the use of a finer strainer (such as woven mesh) to exclude the smaller water creatures. Both types will need to be cleaned periodically, although the strainer box is, to some extent, self-cleaning, since debris is likely to fall off the mesh when water is not flowing into the box (especially if the surface water is disturbed by wind).

Instead of drawing water straight from the pond, it is much better to collect it after it has passed through existing sandy soil (Method 2), or through **sand filters constructed in/on the bed or bank** of the pond (Method 3). Over a period of time, filters in/on the bed are likely to become blocked due to the accumulation of settled suspended solids in and on the filter. Bed filters can only be renovated if the pond is drained to allow their partial or complete reconstruction. It is possible to construct a sloping sand filter down the side of the bank to the pond (Figure 4), but this is rare. With such a filter, some of the filter media can be cleaned or replaced when the water level in the pond drops substantially at some stage.

Man-made filters on the banks (such as horizontal roughing filters and slow sand filters) **or at home** (Methods 4 and 5) can be used to improve the quality of the pond water considerably, and good, well-maintained designs can remove all faecal bacteria, and most viruses. Such filters can be drained down to allow for the regular cleaning of the sand or gravel, or for it to be replaced. A well-trained, dedicated caretaker is needed to supervise the proper running and cleaning of such filters.

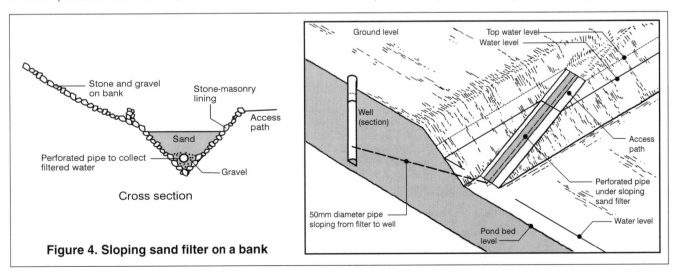

Figure 4. Sloping sand filter on a bank

Improving pond water

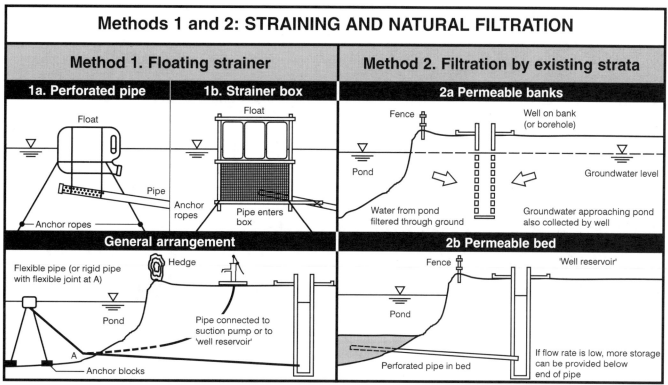

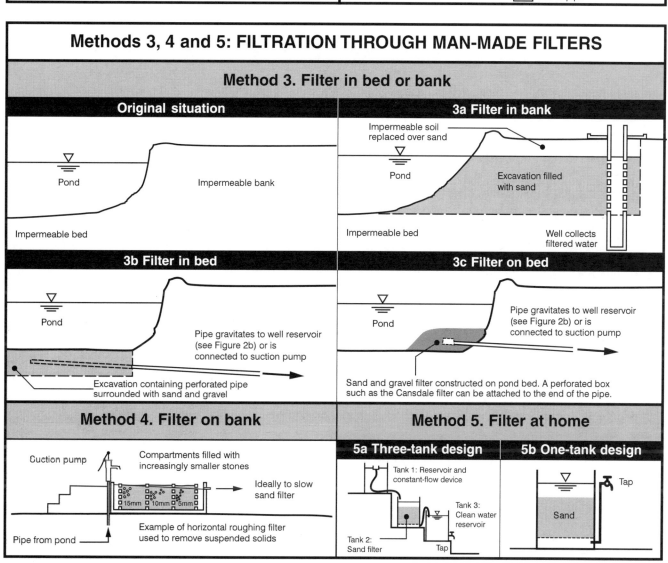

Improving pond water

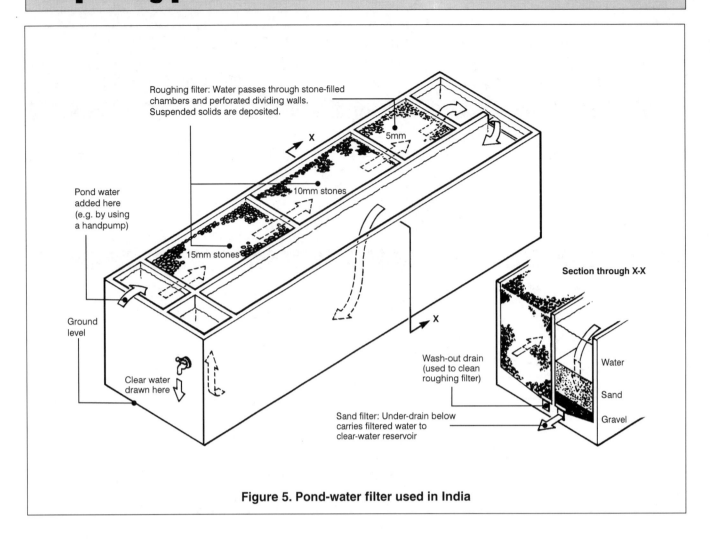

Figure 5. Pond-water filter used in India

Other methods of water purification

Other small-scale water purification methods include **settlement and/or storage** for at least 48 hours (see 'three-pot system' on page 102), **boiling**, **chlorination** and **solar disinfection**.

Further reading

AIIHPH (1993), *Proceedings of the Workshop on Sanitary Protection and Upgradation of Traditional Surface Water Sources for Domestic Consumption*, All India Institute of Hygiene and Public Health, Calcutta.

IRC (1988), *Community Self-Improvements in Water Supply and Sanitation*, Training Series No. 5, International Water and Sanitation Centre (IRC), The Hague.

Pickford, J.A. (1991) (ed.), *The Worth of Water: Technical briefs on health, water and sanitation,* IT Publications, London.

Prepared by Brian Skinner and Rod Shaw

WEDC Loughborough University Leicestershire LE11 3TU UK
www.lboro.ac.uk/departments/cv/wedc/   wedc@lboro.ac.uk

48. Small earth dams

This Technical Brief is concerned with the typical small dam (up to about three metres high) which is built across a stream to form a reservoir. It provides guidance on planning, design and construction, but professional help should always be sought before building any dam whose failure could endanger lives, property or the environment. Care must also be taken to avoid the health hazards of reservoirs, including schistosomiasis and polluted water; and the rights of existing users of the water and land must be protected.

A reservoir is useful where the available flow in the stream is sometimes less than the flow required for water supply or irrigation, and water can be stored from a time when there is surplus, for example, from a wet season to a dry season. In addition to the simple earth dam, alternatives to consider are using the sub-surface (groundwater) dam (see *The Worth of Water,* pages 97-100) or using wells. These may be preferable for environmental and water-quality reasons.

Simple earth dams can be built where there is an impervious foundation, such as unfissured rock, or a clay subsoil. The channel upstream should preferably have a gentle slope, to give a large reservoir for a given height of dam. An ideal dam site is where the valley narrows, to reduce the width of the dam.

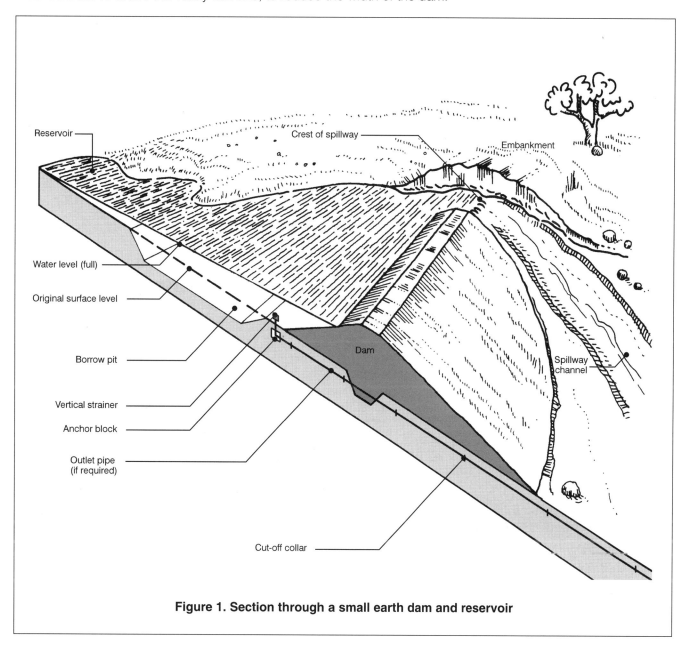

Figure 1. Section through a small earth dam and reservoir

Small earth dams

Design

The design below is suitable for dams up to 3m high. It is a uniform embankment of inorganic, clay loam soil, such as sandy clay loam, clay loam, silty clay loam, or soil with a higher clay content (sandy clay, clay, or silty clay). Any of these can be used provided cracks do not form. The dam must have a 'cut-off' which locks it into the subsoil foundation, ensuring that the dam is stable.

A 3m high dam would typically have a 2m maximum depth of water when full, increasing to 2.5m under flood conditions, with a 0.5m depth of flow over the spillway. The top 0.5m (minimum) is required to provide a safety margin (freeboard) which allows for water rising on the dam due to wind and waves, and wear and tear on the dam crest. The total design height of the dam must be increased for construction by at least 10 per cent, to take account of settlement.

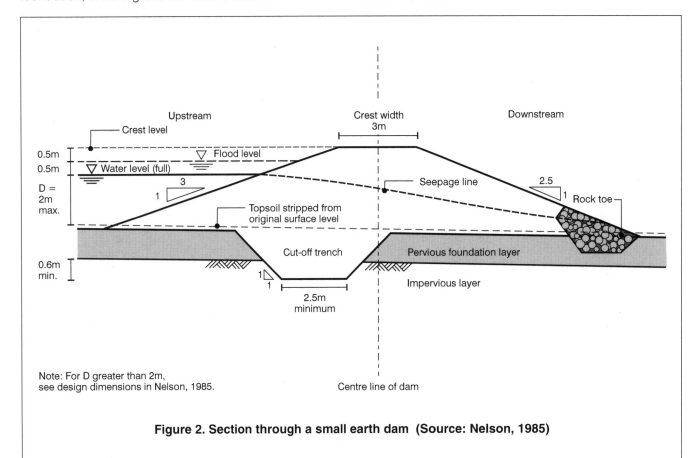

Figure 2. Section through a small earth dam (Source: Nelson, 1985)

Calculating the height of the dam

The height of the dam will depend on the storage required in the reservoir. To calculate this:

- determine the water requirement per day (R litres per day);

- estimate the area of the reservoir (A m^2), the evaporation and seepage losses per day (E mm per day) and, hence, the volume of losses per day (A x E litres per day);

- estimate the length of the critical period (T days), during which the stream flow is less than the water requirement and losses, when requirements would be met using the storage in the reservoir;

- estimate the average stream flow during the critical period (Q litres per day);

- the effective storage required (S litres) = (water requirement per day plus evaporation and seepage losses per day minus average inflow per day) multiplied by the length of critical period:

$$S = (R + A \times E - Q) \times T$$

The dam must be high enough to store this quantity of water. The storage capacity of the reservoir (C litres) is best determined from cross-section surveys across the valley, but can be estimated from the area of the reservoir (A m^2) and the maximum depth of water at the dam (D m) when full:

$$C = 330 \: A \times D$$

The site should then be surveyed to estimate the area (A) of the reservoir for different values of D, and a trial-and-error method will then give the reservoir capacity (C) which meets the storage required (S) and provides a safety margin. The resulting value of A should then be used in the calculation of S to obtain a consistent result. Height of dam = D + 1m.

Small earth dams

Construction

The materials should preferably be taken from the reservoir area; different parts of the side of the valley should be examined so that the most suitable soils are located (soil textures will vary according to position in the valley). The following materials should be avoided: organic material — including topsoil — decomposing material, material with high mica content, calcitic clays, fine silts, schists and shales, cracking clays, and sodic soils. Avoid material with roots or stones.

Other construction points to consider:

- Construct during the dry season.

- Divert the stream; block it with a temporary low dam, or divert it through a culvert (which could become part of the outlet works or spillway later).

- Strip topsoil because it contains organic matter (such as roots) which prevents proper compaction and may provide seepage routes (piping) once the organic matter has decayed.

- Pay attention to people's safety — avoid hazardous practices and dangerous equipment.

- Place material in the dam:

    i) in layers 100 to 200mm deep;

    ii) at the optimum moisture content — when material can be rolled to pencil thickness without breaking, and is as wet as possible without clogging the roller; then

    iii) compact with a heavy roller, or by driving across vehicles or animals.

- Cover the whole dam with topsoil:

    i) plant strong grass (such as Kikuyu grass, star grass or Bermuda grass) to protect against erosion;

    ii) maintain the grass (water in the dry season if necessary), but prevent trees taking root, and keep out animals such as rats and termites.

- Protect the upstream slope:

    i) lay a stone or brush mattress (for example bundles of saplings between 25 and 50mm long) on the slope, and tie it down with wire anchored to posts;

    ii) secure a floating timber beam 2 m from the dam — these need replacing every 10 years or so.

Settlement

Even with compaction, earth dams settle as the weight forces air and water from voids (consolidation) — allow for this settlement in the design.

For small dams, well-compacted settlement should be between 5 to 10 per cent of the height of the dam.

Seepage/filter

Some water will seep through the dam, even if it is constructed of good materials, and well-compacted. This seepage reduces the strength of the dam. Nelson recommends the crest width and slopes shown in Figure 2 to provide a stable, 3m-high embankment making extra seepage protection unnecessary. A safer, but technically difficult, solution is to include a rock toe drain (as shown), to collect seepage water. This should extend up to a third of the height of the dam, and a graded sand and gravel filter must be placed between the dam fill material and the drain to prevent fine clay particles being washed out. The filter must be designed according to the particle size of the dam material and the drain, following, for example, recommendations in Schwab *et al,* p488-490.

Extraction of water from the reservoir

A gravity outlet can be constructed, as shown in Figure 1, using a screened inlet on the bed of the reservoir, and a pipe in a trench below the dam. Problems can arise with seepage through poorly compacted material beside the pipe (reduced by placing seepage collars along the pipe to increase the perimeter by at least 25 per cent), and difficulty repairing a damaged pipe. Alternatively, water can be extracted by lifting or pumping, using some of the methods described in Technical Briefs Nos. 22 and 47, for example:

- a sump (well reservoir) in natural ground at the side of the reservoir, supplied by gravity from a screened inlet and pipe through the bed and side of the reservoir;

- a bank-mounted motorized or human-powered pump; or

- a floating intake.

Safety and management

National and local regulations on small dams must be checked and followed in design, construction, and maintenance.

A technically competent person (an engineer or technician) should be responsible for designing and supervising the construction of the dam. The level of expertise required will depend on the potential for failure. Particular technical attention should be paid to the selection of materials and the design of the filter and spillway.

The sizeing of the spillway is important for protecting the dam during floods, but it is difficult to design. It depends on the rainfall intensity and the size and characteristics of the catchment area, and technical advice should be sought on local standards and practice.

A system needs to be set up for checking the condition of the dam and spillway, and for arranging any necessary repair work. This will usually involve training a local caretaker, who has access to a technician who inspects the dam at an appropriate interval (e.g. before each rainy season).

The dam should be regularly inspected for signs of deterioration, such as cracks, gullies, damage by rodents or insects, seepage, and damage to structures, especially the spillway.

Small earth dams

Spillways

A spillway is required to protect the dam from overtopping, for example during high flows. It passes surplus water downstream safely, preventing both the failure of the dam, and damage downstream.

Surplus water flows over a spillway crest at the top water level, and into an open channel around the side of the dam, discharging safely into the stream below the dam. It may be made from reinforced concrete, but a cheaper solution is a grassed spillway with a:

- vegetated earth channel
- protected crest at reservoir top-water level
- maximum velocity 2.5m/s

A grassed spillway requires regular inspection and maintenance, so that erosion can be repaired and a good grass cover is maintained. It is often used together with a trickle-pipe spillway so that small inflows into a full reservoir flow through the trickle pipe, and do not erode the grass spillway. Table 1 can be used to find the minimum inlet width for a given flood flow. These widths apply to well-grassed spillways. Poorly grassed spillways should be wider.

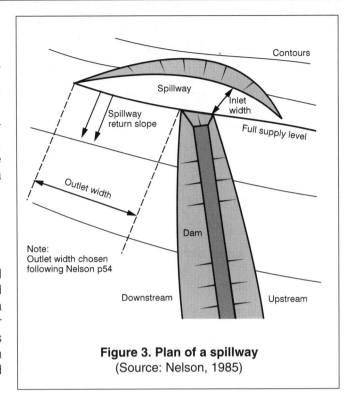

Figure 3. Plan of a spillway
(Source: Nelson, 1985)

Table 1. Minimum inlet width of the spillway

Flood flow (m³/s)	Inlet width (m)
Up to 3	5.5
4	7.5
5	9.0
6	11.0
7	12.5
8	14.5
9	16.5
10	18.5
11	20.0
12	22.0
13	23.5
14	25.5
15	27.5

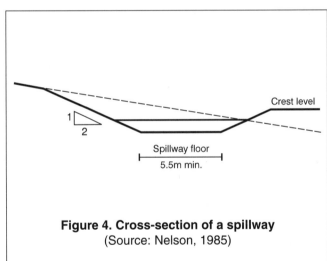

Figure 4. Cross-section of a spillway
(Source: Nelson, 1985)

Further reading

Fowler, John P., 'The design and construction of small earth dams', *Appropriate Technology*, Vol.3, No.4 (reprinted in *Community Water Development*, IT Publications, London, 1989).
Nelson, K. D., *Design and Construction of Small Earth Dams*, Inkata, Melbourne, 1985.
Pickford, John (ed.), *The Worth of Water: Technical Briefs on health, water and sanitation*, IT Publications, London, 1991.
Schwab, G.O., Fangmeier, D.D., Elliot, W.J. and Frevert, R.K., *Soil and Water Conservation Engineering*, Wiley, London, 1993.
Stephens, Tim, *Handbook on Small Earth Dams and Weirs*, Cranfield Press, Bedford, 1991.

Prepared by Ian Smout and Rod Shaw

WEDC Loughborough University Leicestershire LE11 3TU UK
www.lboro.ac.uk/departments/cv/wedc/   wedc@lboro.ac.uk

49. Choosing an appropriate technology

The initial selection of an appropriate technology from a range of possibilities is the key to the successful operation of any facility — technologies are unlikely to function adequately if inappropriate choices are made at the outset. Although this is understood by many, people often underestimate how difficult the choice can be.

This technical brief is intended as a guide to selecting the most appropriate options, taking as its example the selection of water-treatment facilities. It should *not* be seen as a guide for the design of individual treatment processes.

The process contained in this brief can be applied to the selection of single treatment facilities, or as a guide to the development of a *strategy* for a whole area. The process can be used both by people with direct responsibility for making the decisions, and also by other parties to ensure that the right issues are addressed by the decision-makers.

The method described below can be applied to any decision-making process — for example, to identify the technologies for a development project, such as a water supply, sanitation, or refuse-collection scheme.

Case study: water treatment (see, for example, Schulz and Okun, 1984)

There are many different types of water-treatment process to choose from. Table 1 describes some of the more common. Many of the treatment processes used in the South, however, do not work properly.

Table 1. Common water-treatment processes

Water-treatment process	Description	Comments
Plain sedimentation	Allows settlement of heavier particles, which may include much of the solid pollution, and animal (including some very small pathogenic (disease-causing) material).	Very simple and requires no power or chemicals. At its most basic, can be simple storage.
Roughing filter	Filtration though a coarse bed of gravel or coarse sand.	Good for removal of major solid particles and for highly turbid waters.
Slow sand filter	Filtration through a sand bed. Slow flows through the bed ensuring the build-up of a biological layer on the surface of the sand which is an essential part of process.	Removes solid materials and pathogens effectively. Simple to operate. Requires no backwashing of sand to clean — usually only drainage of water and scraping off top biological layer when filtration rate is too slow.
Rapid sand filter	Faster filtration rates through a sand bed — does not have biological-growth layer.	Removal of pathogens not as good as slow sand filter. Requires cleaning by backwashing — passing of water up through filter to remove solid particles that are blocking the flow. Sometimes cleaning by air scouring is also necessary.
Aeration	Water aerated, usually by artificial means — a mechanical device in the tank agitates water — or by spraying.	Good for removal of certain pollutants such as iron and manganese. Requires power.
Coagulation	Addition of chemicals such as alum or lime to bring out pollutants in water — pollutants stick to chemicals and fall to bottom of the tank when allowed to settle.	Requires chemical and power input and control.
Disinfection	Addition of chemicals such as chlorine to kill off disease-causing organisms.	Requires chemical input and control.

Choosing an appropriate technology

Water-treatment process selection

The problem is that many of the treatment processes are *inappropriate* for their use and/or their location. For example, many were developed in the cooler climates of the North, making direct transfer to tropical climates unsuitable. The spare parts, maintenance, and power consumption required by many treatment processes makes them unrealistic options for many parts of the world.

All locations are unique; what is required is not a common *solution* to a problem, but a methodology for the *analysis* of problems.

Figure 1 shows three stages for the selection procedure.

Stage 1: Objectives

The *purpose* of the treatment process must be established. What are you trying to achieve, and why? Is it achievable, is it a realistic goal, is it the main problem? There may be a need to *prioritize* the problems. This stage is often underestimated or taken for granted. For example, in the case of water treatment, the priority in developing countries often should be a low-cost, low-maintenance system.

Stage 2: Analysis

The *constraints* on the proposed development have to be identified and this can only be done by looking at the particularities of the individual case. Often, physical constraints such as water resources and land availability will be taken into account, but other fundamental factors which contribute to the success or failure of a scheme are not adequately addressed.

For analysis purposes we can group the issues to be addressed into the **'SHTEFIE'** criteria, developed at WEDC by Richard Franceys, Margaret Ince and others as a tool to help with analysis of development programmes.

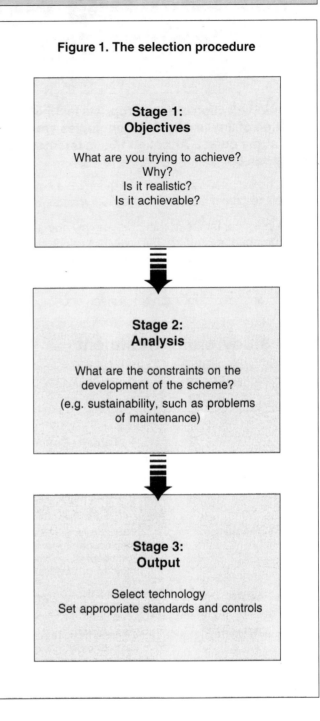

Figure 1. The selection procedure

Stage 1: Objectives

What are you trying to achieve?
Why?
Is it realistic?
Is it achievable?

Stage 2: Analysis

What are the constraints on the development of the scheme?
(e.g. sustainability, such as problems of maintenance)

Stage 3: Output

Select technology
Set appropriate standards and controls

S — SOCIAL
H — HEALTH
T — TECHNOLOGICAL
E — ECONOMIC
F — FINANCIAL
I — INSTITUTIONAL
E — ENVIRONMENTAL

From these groupings, a checklist of factors to consider may be drawn up. Table 2 shows an example of the sorts of issues that could be used for selecting water-treatment options (you should be able to draw up one of your own for your own situation). Think of it as a thought-provoker to ensure that all the relevant factors are taken into account; the SHTEFIE criteria can be useful in this context.

Stage 3: Output

After all the relevant issues have been addressed, the *output* can be evaluated. In the case of water-treatment and most other processes, it is important to realize that there will be *two* main outputs. The first is the technological options themselves. But these are dependent on the methods used to implement and regulate them — usually the water-quality standards set. Often, unrealistic and unattainable standards are laid down with the severely detrimental effect of encouraging people to invest in technologies which are inappropriate for their community. *Options* and *targets/standards* must be considered together, therefore; only then will *appropriate technologies* be selected.

Choosing an appropriate technology

Table 2. The 'SHTEFIE' criteria

S — SOCIAL factors
- Housing facilities; type, distribution
- Public desires and preferences; aesthetic considerations, pressure groups
- Welfare and equity considerations
- Willingness to pay; ability to pay
- Water tariffs, methods and ability to pay
- Population distribution (age, location; growth rates)
- Migration, urbanization
- Cultural and religious aspects, including attitude towards water recycling and sanitation practices
  All of above may affect consumption
- Education levels; structure of workforce; training
- Influence ability to operate and maintain

H — HEALTH factors
- Health statistics, morbidity and mortality rates
- Significant faecal-oral (and other) diseases
- Health services available

T — TECHNOLOGICAL factors
- Water demand and use
- Availability of spare parts and materials
- Availability of local knowledge and expertise
- Present water supply and sanitation facilities; proposed future investments
- Design life of treatment facilities
- Raw water characteristics: source, quantity, quality, availability, and reliability
- Water-quality standards
- Power requirements

E — ECONOMIC factors
- Structure of economy, output by group, industrial and agricultural component
- Major employment sectors
- Foreign-exchange earnings (exports)
  All affect priorities for water supply
- Size of economy, future prospects, balance of payments, trade relations, isolation of economy and vulnerability, distribution of incomes
  All affect ability to pay

F — FINANCIAL factors
- Finance available; method of funding
- Ability and willingness to pay

I — INSTITUTIONAL factors
- Existing roles and responsibilities for organization and management
- Relationships between organizations
- Legislation, policing and regulations

E — ENVIRONMENTAL factors
- Climate, rainfall, hydrology
- Soil conditions, geology, groundwater characteristics
- Water-resource availability
- Impact of any plant: noise, smell, insects, visual impact, health considerations
- Sustainability

Example

A village currently obtains its water from a large stream, source 1, about 50m away. Although the stream water is highly turbid, and the incidence of diarrhoeal disease is high, villagers are used to obtaining their water from this supply because it is close by.

The local health worker has recommended that a 'cleaner' water supply should be sought, as she believes that this would improve the health of the villagers significantly. There is a second water source on the other side of the village, about 500m away (source 2). This source is not used very often because of the distance, but it appears to be much less turbid than the first source.

What would you recommend as a suitable course of action?

Note that the example given here is simplified. In normal situations more factors will usually have to be considered. It is used only to illustrate how the selection process may be applied.

Stage 1: Objectives

To improve the health of the villagers. Requires a cleaner water supply, probably requiring treatment to remove pathogens — but all at an acceptable cost.

Choosing an appropriate technology

Stage 2: Analysis — using SHTEFIE

Factor	Effect	Outcome
Social Social desires for increased convenience of supply mean that many villagers want a piped water supply to standposts — to be of better quality than the existing supply	New water supply must appear better than old or people will not accept it	Removal of turbidity is a priority so new source appears to be better — sedimentation, filtration or coagulation may be suitable
Health High incidence of diarrhoeal diseases in village	Pathogen removal required	Process needs to include pathogen removal — possibly slow sand filtration and/or chlorination
Technological Chemicals and spare parts difficult to obtain	Process to be simple and not reliant on chemicals or power	Rules out coagulation and possibly chlorination. Also rules out rapid sand filters
Economic Large amount of agriculture in area requiring irrigation water	Large amount of reasonable (not drinking-water standard) quality water required	Possibility of using higher-quality source for drinking-water supply and lower-quality source for irrigation
Financial Income levels still low in village	Ability to pay is low	Need for inexpensive options
Institutional There is little involvement of water agencies in the area	Operation and maintenance capabilities are likely to be low	Need for simple options
Environmental Rainfall is fairly even throughout the year	Need to check reliability of flow in streams	May need to have further source of water supply if flow is low.

Stage 3: Output

Of the main treatment options listed in Table 1, the analysis has revealed that sedimentation and slow sand filtration are probably the most appropriate treatment options because of the operational and maintenance requirements. Chlorination could be considered if completely safe drinking-water were required, but the chemical requirement might mean that this option is not appropriate. Water from source 2 could be used for drinking-water supplies after treatment, leaving the water from source 1 for irrigation purposes. Otherwise, the very high turbidity in water source 1 would mean that a pre-treatment stage such as roughing filtration may have to be employed. Water-quality targets should be to remove turbidity and pathogens to acceptable levels, and to perform the routine operational tasks for the slow sand filter when required. (For further details about the operational requirements of slow sand filters, refer to *The Worth of Water.*)

Conclusions

So, when selecting any technology, consider the following:

Objectives: *What is required? Why? Is it realistic?*
Analysis: *Can it be achieved? What are the limitations?*
Output: *What technologies and controls are appropriate given the problem and the constraints?*

Further reading
Pickford, J. (ed.), *The Worth of Water,* IT Publications, London, 1991.
Shulz, C.R. and Okun, D.A., *Surface Water Treatment for Communities in Developing Countries,* John Wiley & Sons/IT Publications, London, 1984.
WELL, *Guidance Manual on Water Supply and Sanitation Programmes,* WEDC for DFID, Loughborough, 1998.

Prepared by Jeremy Parr and Rod Shaw

WEDC Loughborough University Leicestershire LE11 3TU UK
www.lboro.ac.uk/departments/cv/wedc/   wedc@lboro.ac.uk

50. Sanitary surveying

When visiting water-supply schemes, it is usually possible to spot any faults and deficiencies that could lead to the pollution of potable water. Sanitary surveying is an inspection technique that records such visible problems, enabling fieldworkers to assess the likely quality of the water, relative to other sources. Figure 1 shows a woman collecting water from a stream which could be polluted by human excreta and urine, animal and domestic wastes, soaps and detergents, pesticides and fertilisers.

Sanitary surveying formally identifies possible pollution problems which may threaten drinking-water quality at the source, point of abstraction, treatment works, or distribution system. It relies on the inspection of physical installations by an inspector or a team of inspectors.

Figure 1. Possible causes of water-source pollution

Sanitary surveys can be carried out at any one of the three points of a water-supply scheme (Figure 2):

1. at the source and intake (to assess whether the quality of the raw water is at risk, and whether the abstraction method is satisfactory);

2. at the treatment works (to assess whether suitable treatment processes are being used, and whether correct procedures are being followed); and/or

3. at the distribution system (to assess whether the quality of the water is put at risk during distribution).

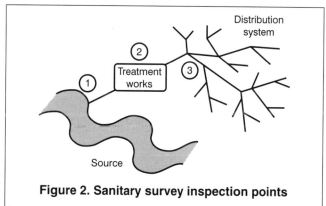

Figure 2. Sanitary survey inspection points

Sanitary surveying

What is the purpose of a sanitary survey?

In carrying out a sanitary survey, an inspector is identifying potential risks to the quality of the water — but she or he should also take the opportunity to make constructive criticism, leading to positive improvements. *It should not be an opportunity to indulge in destructive criticism.*

Undertaking a sanitary survey should also be considered:

- when new water sources are being developed, to assess the water quality and any treatment needs;
- when comparing water sources for potential development;
- when contamination is suspected, to identify the likely cause;
- when there is an epidemic of a water-borne illness, to identify the likely cause;
- to interpret results from water-quality analysis, to establish how the water became contaminated;
- as a routine exercise, to monitor sanitary conditions; or
- when there are significant changes (such as heavy rain or construction activity) which could affect water sources.

Sanitary surveying and water-quality analysis (either in a laboratory or in the field) are complementary activities; they are both important, and both have limitations (see Table 1).

What data is needed for a sanitary survey?

Certain basic data is needed to identify where sanitary surveys are required:

- population data for each town, village, and community;
- information on water sources;
- summaries, from past studies, of data for water quality;
- identification of sources for which no water-quality data is available;
- summaries of health records on the incidence of illnesses associated with water quality and sanitary conditions;
- correlation between outbreaks of illnesses, and water source and quality; and
- any water-treatment methods being used.

Sanitary-risk factors

During a sanitary survey, every insanitary situation that could increase the risk of illness is termed a 'Sanitary-risk factor'. The importance of each risk cannot always be quantified — some risks may be more important than others; some may combine unfavourably — but each risk needs to be eliminated if at all possible.

- Identified sanitary-risk factors are not ranked in order of priority; each risk receives equal weighting.

Table 1. Water-quality analysis and sanitary surveying

Water-quality analysis	*Sanitary surveying*
■ Water-quality analysis is expensive, requires equipment and competent staff and, therefore, is not always easy to perform regularly or routinely.	■ Sanitary surveying is cheap, requires neither equipment nor highly-skilled staff, and may easily be performed regularly or routinely.
■ Water-quality analysis provides only a snapshot — a record of the water quality at the time of sampling.	■ Sanitary surveying can reveal conditions or practices that may cause isolated pollution incidents or longer-term pollution.
■ Water-quality analysis will indicate whether a water supply is contaminated; but, usually, will not identify the source of contamination.	■ Sanitary surveying reveals the most obvious possible sources of contamination, but may not reveal all sources of contamination, for example, remote contamination of groundwater.

Sanitary surveying

- Important, potential sanitary-risk factors — there are usually about ten — should be identified. Equal ranking enables a sanitary-risk score (from 0 (no risk), to 10 (very high risk)) to be established, based on a sanitary survey. The use of 10 sanitary-risk factors (a standard number) makes comparing different sources easy. By using 10 sanitary-risk factors for each source, some risks may be omitted from lists. Some risks may be on-site (local); others may be off-site (remote/distant).

The following categories of sanitary-risk score are frequently used:

Sanitary risk score	Assessment of risk
> 9	Very high
6, 7, 8	High
3, 4, 5	Moderate
0, 1, 2	Low

The reasons for the presence of sanitary-risk factors at water-supply schemes may be attributed to any of the following:

- poor site selection;
- poor protection of the water-supply scheme against pollution;
- inappropriate construction;
- structural deterioration or damage; and/or
- lack of hygiene knowledge/education of users or local inhabitants.

Who should undertake sanitary surveys?

The training and experience that inspectors require to be able to undertake sanitary surveys depends on the size of the population, but all inspectors should have a basic knowledge and understanding of water-supply technology, public-health principles, water-supply operations, and management. A shortage of experienced staff should not prevent sanitary surveys being undertaken, although simple training programmes may be needed.

Personal qualities are very important. Inspectors should be thorough, professional, conscientious, honest, and constructive; what is learned will depend on how thorough and perceptive the inspector is.

The elimination of certain sanitary-risk factors might be difficult. Major repairs or improvements or identifying suitable water-treatment processes may require specialist assistance.

Illustrated sanitary report forms

Sanitary-survey reports should be done quickly, and be simple and accurate. This is straightforward if illustrated report forms are used. Figure 3 shows a report form for a hand-dug well — and similar documents can be prepared or adapted for other water sources and situations.

On one side of the form there is a schematic illustration of the water source and abstraction point, the treatment process, or distribution system. Possible sanitary-risk factors are identified by numbers, which correspond to the questions opposite. Each question should be phrased in such a way that a 'Yes' answer indicates a sanitary-risk factor.

By using illustrated report forms, inspection teams can:

- identify possible points of contamination for a water source or supply scheme;
- quantify the level of risk for each water source or supply scheme;
- provide a visual illustration of where there are risks, and why; and
- retain a clear record — providing guidance for the user of the remedial work needed.

How are the sanitary-survey results used?

One copy of the inspection form should be handed to the user, and a second copy filed/stored. Prior to a sanitary survey, the inspector should study the past inspection forms for each water source. A sanitary survey will only be fully effective if action is taken to eliminate the sanitary-risk factors identified. All interested parties (water-quality agencies, water-supply agencies, etc.) should be informed of any necessary improvements.

Sanitary surveys of treatment plants should be conducted regularly (at least once a year) or when evidence suggests that they are necessary.

If water quality is found to be unsatisfactory, take the following action:

- repeat the analysis of water samples from the affected area to check the reliability of the initial unsatisfactory findings;
- carry out a sanitary survey;
- carry out a more detailed investigation of the source, intake, treatment works and distribution system;
- carry out remedial repairs, construction work or improvements to remove the sanitary-risk factors identified; and
- repeat the analysis of water samples from the affected area to check whether remedial work has been successful.

Sanitary surveying

Sanitary-survey form for assessment of risks for contamination of a hand-dug well

A. General information

Location of hand-dug well:
- Village: ....
- Location within village: ....
- Identification reference: ....

Date of visit: ....

Was a water sample taken? Yes / No

Sample reference ....

Total score of risks ....../12

Sanitary risk score:	9, 10, 11, 12	=	very high
	6, 7, 8	=	high
	3, 4, 5	=	moderate
	0, 1, 2	=	low

Signatures

Community representative: ....

Inspector: ....

B. Identification of sanitary-risk factors — Yes / No

1. Is there a latrine within 10m of the well? ☐ ☐
2. Is the nearest latrine on higher ground than the well? ☐ ☐
3. Is there any other source of pollution (e.g. animal excreta, rubbish) within 10m of the well? ☐ ☐
4. Are the rope and bucket exposed to contamination? ☐ ☐
5. Is the height of the headwall (parapet) around the well inadequate? ☐ ☐
6. Is the headwall (parapet) around the well cracked or broken? ☐ ☐
7. Is the concrete apron around the well less than 1m wide? ☐ ☐
8. Is there poor drainage, allowing stagnant water within 2m of the well? ☐ ☐
9. Is the concrete apron around the well cracked? ☐ ☐
10. Are the walls of the well (well-lining) inadequately sealed? ☐ ☐
11. Is the drainage channel cracked or broken, allowing ponding? ☐ ☐
12. Is the fencing around the well inadequate to keep animals away? ☐ ☐

Figure 3. An example of an illustrated sanitary report form

Note: Illustrations may not be comprehensive. They may need adaptation, and should not be a substitute for thinking!

Further reading

Lloyd, B. and Helmer, R., *Surveillance of Drinking-Water Quality in Rural Areas*, Longman, Harlow, 1991.
McNeill, D., *Manual for the Appraisal of Rural Water Supplies*, ODA, London, 1984.
Hofkes, E.H., (ed.), *Small Community Water Supplies*, IRC Technical Paper 18, IRC, The Hague, 1986.

Prepared by Michael Smith and Rod Shaw

WEDC Loughborough University Leicestershire LE11 3TU UK
www.lboro.ac.uk/departments/cv/wedc/   wedc@lboro.ac.uk

51. Water, sanitation and hygiene understanding

The degree to which many illnesses spread is related to people's living environment. Real, sustainable improvements require an understanding of the problems that a poor environment causes, and the benefits that cleaning up the environment can bring. This Technical Brief outlines the main issues of hygiene understanding and its important role in water and sanitation projects.

Factors that can increase disease:

Poor sanitation
- lack of appropriate and well-maintained excreta-disposal facilities
- lack of refuse collection
- inadequate control of vectors

Poor water
- limited quantity of water for hygiene purposes
- poor-quality water

Poor knowledge and practice
- low level of hygiene understanding
- poor hygiene practice (e.g. food contamination from soiled hands)

Poor housing and drainage
- poor, overcrowded housing
- inadequate drainage systems

Figure 1 shows the causes and transmission routes of environmental-related illnesses. Appropriate water supplies, sanitation and good hygiene practice not only improve health, but may also bring other, secondary, benefits:

- improved agricultural practices and nutrition
- greater chances to enhance socio-economic development
- increased standard of living and convenience

Technical solutions on their own are not enough. *They must be correctly operated and maintained, and there must be the will, financial capability and understanding within the community to manage such solutions.* Education, therefore, is a key activity in any attempt to improve health, and must be based on a 'people-centred' approach for maximum benefit. Before undertaking a new water or sanitation project, present activities, people's everyday behaviour and their knowledge need to be understood and considered.

Water

A sufficient quantity of water — of an acceptable standard — is a prerequisite to life itself. Bringing supplies nearer to the home can save time for those, mainly women, who trek long distances to collect water.

But water needs to be *properly managed* in order to provide the greatest benefit.

Examples of water management tasks include:

- protection of sources and supplies
- operation and maintenance of water and sanitation facilities
- drainage
- wastewater disposal

Increasing water quantity for people who have a good understanding of hygiene and put their knowledge into practice, will have a greater impact on general health than an improvement in water quality on its own.

Technical Brief No. 52 will discuss the issue of water quality and quantity in more detail.

Sanitation

Sanitation is a measure that is undertaken to protect health. The three main categories of sanitation are:

- excreta disposal
- refuse disposal
- vector control

Excreta disposal

The appropriate disposal of excreta is one of the most effective barriers to disease transmission. Faeces contain many pathogens (disease-causing organisms) and can also contain parasites (organisms that live in a host such as a human being). Both can cause illness, but this can be prevented if faeces are disposed of correctly. Methods of appropriate disposal of excreta include:

- pit latrines
- septic tanks with soakaway fields
- sewerage and wastewater-treatment facilities

In many developing-country situations, the pit latrine is the most appropriate method of excreta disposal as it is simple, easy to build and operate and maintain. There are many different pit latrine arrangements, each of which is suited to different situations. Variations include:

- dry or wet pit (pour flush)
- simple or ventilated improved
- single pit or double pit
- individual or communal
- pit or borehole, etc.

Water, sanitation and hygiene understanding

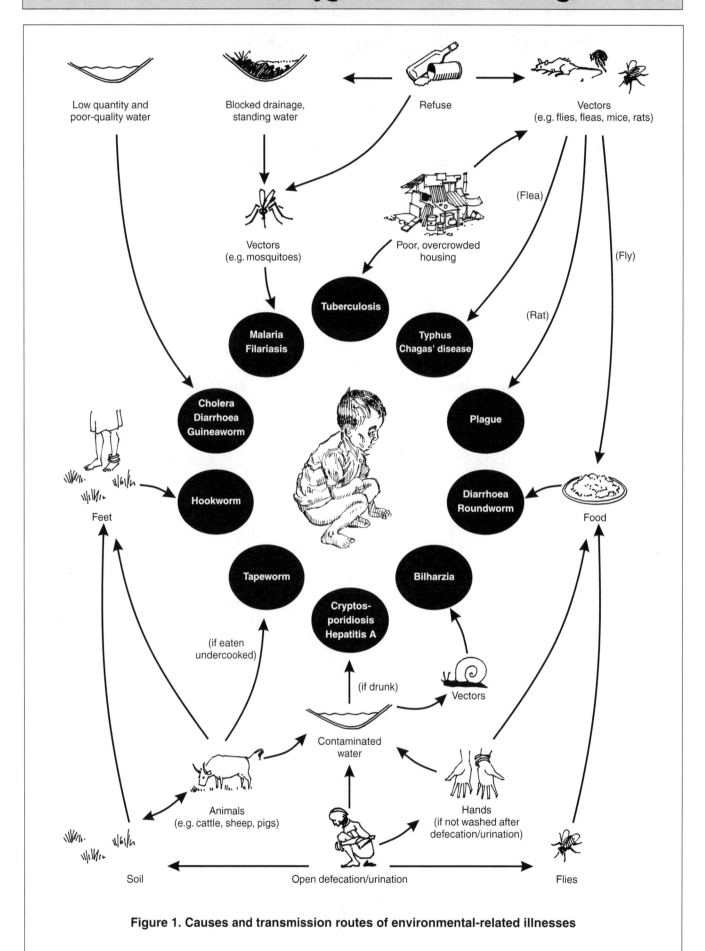

Figure 1. Causes and transmission routes of environmental-related illnesses

Water, sanitation and hygiene understanding

Refuse disposal

Refuse that is not disposed of in a sanitary manner can become a breeding or feeding place for vectors, and can cause an increase in the spread of disease. Appropriate disposal includes:

- recycling of valuable materials
- re-use of organic materials as fertilisers
- burying in pits in the ground
- incineration

Vector control

Vectors such as rats, fleas, flies and mosquitoes can all transmit disease. Methods to reduce vector numbers include:

- improving excreta-disposal methods
- improving refuse-disposal facilities
- improving drainage to remove standing water
- chemical and biological methods of control

Understanding hygiene

If communities are to benefit from the technology designed to improve their health, people have to understand the basics of hygiene and its role in disease prevention.

How should hygiene education be undertaken?

Hygiene education should not be authoritarian, with one-way communication. It should be people-centred with, at least, two-way, or at best, multi-way communication as shown in Figure 2 (Linney, 1995).

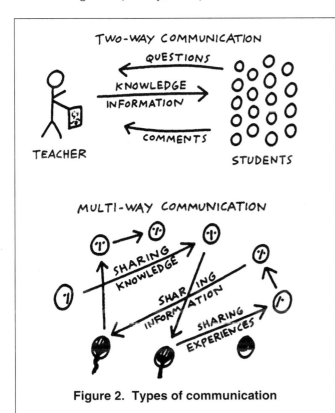

Figure 2. Types of communication

Examples of good hygiene practice must be clearly understood if they are to be effective. They have to relate to the person's life situation and be perceived to bring measurable benefits to that person's life so that the person will want to practise what he or she has learned. It is better to develop people's existing understanding of hygiene and the causes of illness than to have an authoritarian approach imposing totally new ideas and ridiculing existing ideas.

Tools for hygiene education

One of the challenges of hygiene education is how to reach everybody in the community. Using a variety of tools can help to solve this problem.

Suitable tools include visual aids such as:

- pictures (preferably drawn by somebody from the community concerned)
- drama / theatre groups
- television and radio

Using drama or theatre groups can attract wide-ranging audiences; having such events in market-places or open spaces attracts a lot of attention.

Where should hygiene education take place?

Places where people gather are ideal settings for hygiene education programmes, as large numbers from both sexes and a range of ages, class and social status can be reached. Typical examples are:

- schools
- health centres and clinics
- places of worship
- market-places or open spaces
- community meetings

Examples of good practice

- using clean vessels to collect and store water;
- covering storage containers;
- separating drinking-water containers from other water containers (e.g. bowls used for hand-washing, cooking pots, containers used for watering crops);
- keeping areas for collecting and storing clean water free from waste and standing water;
- cleaning latrines regularly;
- disposing of all excreta properly, preferably in a latrine;
- washing hands after excreting, and before preparing food, using soap where possible;
- covering food;
- using clean cooking utensils; and
- disposing of wastewater and refuse in a pit (or as appropriate to the community concerned).

Water, sanitation and hygiene understanding

Examples of good hygiene practice should be *locally formulated*, and care must be taken to ensure that hygiene is not considered inappropriate and an 'unaffordable luxury'. What people know and what they do are often two very different things. It should be noted that water, sanitation and hygiene education may not always be perceived as being high priorities in terms of people's income and time. Obtaining food, shelter, money and clothing may all be higher on a person's list of priorities than obtaining high-quality water and sanitation. Hygiene education programmes have to take this into account, providing explanations of the benefits of improved hygiene such as improved health for the family. This in turn will result in:

- less time off work;
- less need to buy medicines;
- less likelihood of premature deaths in the family; and
- economic benefits.

It could also mean that:

- people's surroundings are more pleasant if refuse and standing water are reduced;
- offensive smells are reduced if excreta is disposed of appropriately;
- food suffers less damage from rats or mice if properly stored and rats and mice are less numerous; and
- the family's status will increase if the family has its own latrine.

Innovative promotional and motivational measures are also needed, such as providing incentives to attend hygiene education sessions (i.e. free soap or food). The choice of session leader or promoter can also influence the effectiveness of a programme. Hygiene instruction presented by respected locals is more likely to be taken seriously.

Who should be involved?

Water, sanitation and health education programmes must involve *the whole community* – women, men, and children of all ages, classes and social status. Full involvement in the planning, design, implementation and evaluation stages of a project is vital if it is to have lasting benefits. It should also be culture- and gender-sensitive, and take account of the different responsibilities people have for promoting good hygiene practice within the community.

Separate hygiene education sessions for people grouped together according to sex or age can sometimes help to ensure equal participation.

Key points

For water and sanitation projects to have a positive effect on health, the following are also required:

- correct operation and maintenance of water-supply and sanitation systems;
- the will, and financial and managerial capacity within the community to undertake system operation and maintenance;
- community-wide understanding of the importance of hygiene, and the benefits that it can bring; and
- the practise of improved hygiene.

Further reading

Almedom, A., Blumenthal, U. and Mandeson, L., *Hygiene Evaluation Procedures: Approaches and methods for assessing water- and sanitation-related practices*, IT Publications, London, 1997.

Boot, M.T. and Cairncross, S., *Actions Speak: The study of hygiene behaviour in water and sanitation projects*, IRC/London School of Hygiene and Tropical Medicine, London, 1993.

Levert, L., 'Clean up your act — development theatre for water and sanitation', *Waterlines,* 14 (1) 27-31, IT Publications, London, 1995.

Linney, B., *Pictures, People and Power,* Macmillan, London, 1995.

Nilanjana Mukherjee, *People, Water and Sanitation. What they know, believe and do in rural India,* Stambh Design Consultancy, New Delhi, 1990.

Werner, D. and Bower, B., *Helping Health Workers Learn: A book of methods, aids and ideas for instructors at the village level,* The Hesperian Foundation, Palo Alto, 1982.

Wood, S., Sawyer, R. and Simpson-Herbert, M., *PHAST Step-by-Step Guide: A participatory approach for the control of diarrhoeal disease* (WHO/EOS/98.3), WHO, Geneva, 1998.

Prepared by Sarah House, Margaret Ince and Rod Shaw

WEDC Loughborough University Leicestershire LE11 3TU UK
www.lboro.ac.uk/departments/cv/wedc/   wedc@lboro.ac.uk

52. Water — quality or quantity?

Anyone thinking of implementing a water project must clearly understand water quality and quantity requirements. This Technical Brief looks at these requirements and compares their importance in relation to improving people's health.

The previous Technical Brief discussed, in some depth, just how important good hygiene understanding and practice are; improved water quality or quantity alone will not necessarily improve health if communities do not have an understanding of the concepts of hygiene and disease transmission. If positive benefits are to ensue, communities must also have the will, and the financial and management capabilities to be able both to operate and maintain water projects, and to put into practice what they know about hygiene.

Water (or lack of it) can play a part in the transmission of diseases in various ways. The four water-related transmission routes are highlighted in Table 1.

Many of the water-borne, water-based and water-washed diseases are transmitted through the 'faecal-oral' route; pathogens or parasites from the faeces of one person are transmitted by various routes to the mouth of another, and in this way cause illness. Some diseases, however, such as skin or eye infections, diseases caused by lice or mites, or those caused by pathogens or parasites which penetrate the skin, are not transmitted by this route. For these diseases the main prevention strategies are improved hygiene understanding and practice, and reducing contact with the contaminated medium.

Table 1. Disease transmission and preventive strategies
(Adapted from Cairncross *et al.*, 1983.)

Classification	Transmission	Examples	Preventive strategies
Water-borne (water-borne diseases can also be water-washed)	Disease is transmitted by ingestion	· Diarrhoeas (e.g. cholera) · Enteric fevers (e.g. typhoid) · Hepatitis A	· Improve *quality* of drinking water · Prevent casual use of other unimproved sources · Improve sanitation
Water-washed (water scarce)	Transmission is reduced with an increase in water quantity: · Infections of the intestinal tract · skin or eye infections · infections caused by lice or mites	· Diarrhoeas (e.g. amoebic dysentery) · Trachoma · Scabies	· Increase water *quantity* · Improve accessibility and reliability of domestic water supply · Improve hygiene · Improve sanitation
Water-based	The pathogen spends part of its life cycle in an animal which is water-based. The pathogen is transmitted by ingestion or by penetration of the skin.	· Guinea worm · Schistosomiasis	· Decrease need for contact with infected water · Control vector host populations · Improve *quality* of the water (for some types) · Improve sanitation (for some types)
Insect-vector	Spread by insects that breed or bite near water	· Malaria · River blindness	· Improve surface-water management · Destroy insects' breeding sites · Decrease need to visit breeding sites of insects · Use mosquito netting · Use insecticides

77

Water — quality or quantity?

Table 2. Recommended minimum water-quantity requirements	
Usage	Water usage (litres per head per day unless otherwise stated)
Individuals	15 to 25
Schools	15 to 30 litres per pupil per day
Hospitals (with laundry facilities)	220 to 300 litres per bed per day
Clinics	Out-patients 5 In-patients 40 to 60
Mosques	25 to 40
Pour-flush latrines	1 to 2 litres per flush 20 to 30 litres per cubicle per day
Dry latrines (for cleaning)	2 litres per cubicle per day (more if heavy usage such as in refugee camps)
Livestock: large (cattle)	20 to 35
Livestock: small (sheep, pigs)	10 to 25

There are many water uses (e.g. drinking, cooking, washing, agriculture etc.) and the quantity and quality required for each varies. Drinking-water requirements are usually the most stringent.

Basic requirements for drinking-water

- There must be enough to prevent dehydration.
- It should be acceptable to the consumer. (A bad taste or colour, staining, or unpleasant odour can cause a user to choose an alternative source.)
- It should be free from pathogenic (disease-causing) organisms and toxic chemicals.
- It should not cause corrosion or encrustation in a piped water system, or leave deposits.

Table 3. Collection distance implications on water quantity	
Distance to water-point	Water consumption (litres per person per day)
Walking distance > 1000m to communal water-point	5 to 10
Walking distance < 250 m to communal water-point	15 to 50
House or yard connection — single tap	20 to 80

Quantity of water

The minimum quantity of drinking-water needed for survival is three to five litres per person per day depending on the temperature, and an individual's level of exercise. Table 2 gives further details of water-quantity requirements.

The quantities used will fluctuate with distances that have to be walked to collect water (Table 3). It should be expected, therefore, that usage will increase with the improved convenience of a piped supply, when a new source nearer to the home is realized, or when income levels increase (Table 4).

Increased quantity of water can also improve:

- agricultural practices
- nutrition
- socio-economic growth

Quality

Pollutants and the physical features of water can affect health in the following ways:

- some can be directly harmful to health, such as microbiological and biological contaminants, fluoride, pesticides and industrial pollutants;
- colour, taste, turbidity and odour can make the water objectionable to consumers, and cause them to use another, superficially less objectionable, but not necessarily safer, source; and

Water — quality or quantity?

- others such as pH and turbidity can reduce the effectiveness of treatment processes such as disinfection.

Microbiological and biological contaminants are the major source of illness.

The World Health Organization (WHO) has produced guideline levels for quality for use as targets and as an aid for countries who wish to produce their own. In many regions, however, the WHO guideline levels may not be achievable in the short term and, therefore, interim national standards should be set which promote improved water quality and which are realistic. *Setting targets that are too high can be counterproductive; they may be ignored if they are not attainable.* National standards should reflect national conditions, priorities and capacity to improve water supplies, especially in small communities where the choice of source and treatment are limited, and finances are constrained.

E.coli (or thermotolerant coliforms) are used as indicators of faecal pollution. If *E.coli* are present then it is likely that pathogens are also present. The WHO guideline level for thermotolerant coliforms indicates that, for all water intended for drinking, none should be detectable in any 100ml sample. Alternative figures are often quoted which are more appropriate for rural communities and emergency situations (Table 5).

Water-quality data gives information about the present situation but does not show the patterns of intermittent or seasonal pollution. A *sanitary survey* (see pages 69-72) will give information about the likelihood of faecal pollution. *Local knowledge* and *local medical information* can also help in assessing pollution problems.

When making an assessment of drinking-water quality, the investigator should be aware that *drinking-water can often become contaminated from unclean collection vessels or storage containers in the home.*

In general, microbiological pollution levels of sources vary from low levels in rainwater (if it is collected in a clean environment), deep groundwater and springs (unless in an area of highly fissured rock), to high levels in shallow groundwater (unprotected hand-dug wells), rivers, streams and lakes.

Table 4. Economic circumstances and domestic water use
(Adapted from Twort *et al.*, 1994, p7)

Economic circumstances	Quantity of water used for domestic purposes (litres per person per day)
Upper to middle-income groups (warm climate: piped supply to home)	200
Upper to middle-income groups (Europe: piped supply to home)	165
Low-income groups (warm climate: standpipe supply) · urban · rural (washing at standpipe) · rural (drinking and washing only)	70 65 25
Low-income groups (Europe: piped supply to home) · small flat with shower	100

Table 5. Thermotolerant coliform guide
(Adapted from Ockwell, 1986, p327)

Level of faecal pollution (number of thermotolerant coliforms present)	Inference
0 - 10	Reasonable quality
10 - 100	Polluted
100 - 1000	Dangerous
> 1000	Very dangerous

Water — quality or quantity?

Quality can be improved by:

- source protection;
- improved hygiene awareness and practice;
- improved sanitation;
- water treatment;
- efficient and safe distribution to the consumer; and
- good storage practices.

Quality versus quantity

Steven Esrey highlights the relative impact of interventions on the reduction in diarrhoeal diseases (Table 6). From this it can be seen that quantity has a greater effect than quality, and also that good hygiene and sanitation practice have even greater impacts.

Table 6. The effect of interventions on the reduction of diarroheal diseases

Intervention	Reduction in diarrhoea (approx. %)
Water quality	15
Water quantity	20
Hygiene	33
Sanitation	35

Summary

When setting up a water-supply programme, the following points should be noted:

- In general, an increase in water quantity is more beneficial than an increase in water quality.

- The relative importance of water quality and water quantity depends on the situation. In urban areas or in refugee situations, for example, where large numbers of people live in close proximity, greater care must be undertaken to prevent epidemics. The quality of water, therefore, becomes more important.

- An excess supply of water can lead to other health hazards, such as standing water.

- In general, sanitation and hygiene understanding have a greater impact on health than improvements in water quality or quantity.

Further reading

Cairncross, S. and Feachem, R.G., *Environmental Health Engineering in the Tropics: An introductory text,* John Wiley, Chichester, 1983.
Esrey, S.A., 'No half measures — sustaining health from water and sanitation systems', *Waterlines,* Vol.14 No.3, 24-27, IT Publications, London, 1996.
Hofkes, E.H. (Ed.) IRC, *Small Community Water Supplies — Technology of small water supply systems in developing countries*, IRC Technical Paper No.18, IRC Water and Sanitation Centre, The Hague, 1998.
Ockwell, R., *Assisting in Emergencies: A resource handbook for UNICEF field staff,* UNICEF, New York, 1986.
Pickford, J. (Ed.), *The Worth of Water: Technical briefs on health, water and sanitation,* IT Publications, London, 1991.
Twort, A.C., Law F.M., Crowley, F.W. and Ratnayaka, D.D., *Water Supply,* Edward Arnold, London, 1994.
WHO, *Guidelines for Drinking-Water Quality, Vol. 1,* Geneva, 1993.

Prepared by Sarah House, Margaret Ince and Rod Shaw

WEDC Loughborough University Leicestershire LE11 3TU UK
www.lboro.ac.uk/departments/cv/wedc/   wedc@lboro.ac.uk

53. Training

Water and sanitation facilities will only be sustainable if there are enough competent people to plan, construct, operate, maintain and manage them. Training is a critical factor; this Technical Brief looks at key elements in its effective provision.

What is training?

Training is the systematic development of knowledge, skills and attitudes (KSA) required to work effectively. Training aims to change behaviour. It is an agent of change.

For example, water-supply operators with limited skills and knowledge in water treatment can, through training, be made aware of the importance of variations in raw-water quality, and become motivated and skilled to act to ensure the supply of safe drinking-water.

The training process

Knowledge and skill on their own will not lead to changed behaviour unless accompanied by motivation and a supportive environment.

Responsibility for the effectiveness of training is shared by the individual, the organization, and the trainers. The term 'organization' can be interpreted broadly, e.g. a government department, an aid agency, or a community management structure.

The individual
- needs motivation and the ability to take advantage of training

The organization
- ensures training matches needs
- provides suitable climate to motivate trainee
- ensures conditions exist to utilize newly acquired knowledge and skills

The trainers
- provide the opportunities for learning to take place

Trainers may be from within the organization or from an external agency. They do not have to be designated 'trainers' within an organization, but could be managers with motivation and enthusiasm to promote learning.

Training is not an isolated activity

Training is not an isolated activity of instruction. It should be a cyclical process with distinct stages, as shown overleaf.

Training

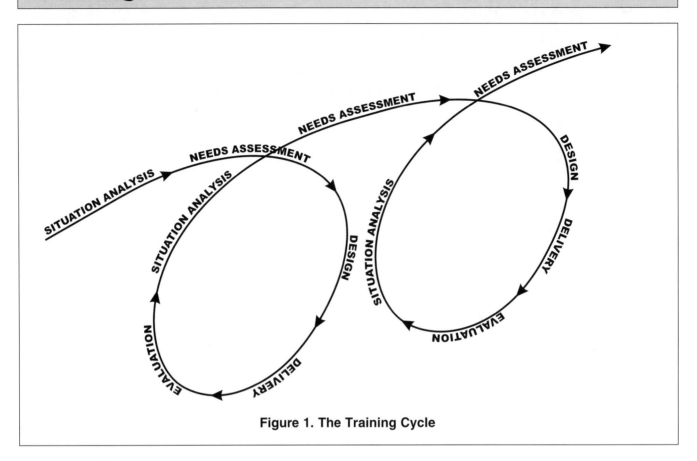

Figure 1. The Training Cycle

Situation analysis

Training should not be for training's sake. There should be clearly identified reasons for training, based on an analysis of a community's or an organization's situation. Do problems exist? Can they be solved by training? A pump caretaker may be well-trained but, without spare parts, she cannot be effective.

Common problems related to work performance, and how training might help, include:

- Increasing knowledge to overcome misunderstandings
- Skills training to overcome a lack in this area
- Supporting individuals, through training, to motivate and develop attitudes

Training-needs assessment

Training should be based on clearly identified needs. Training-needs assessment involves the collection and analysis of information related to the work required. With this information, decisions can be made on:

- *who* needs training
- *how many* people need training
- *what kind* of training is required and to what standard

Assumptions based on occupational labels should be avoided. It is necessary to establish the actual work required to be done in each situation. The label 'pipefitter', for example, is very broad and it may be misleading because it can be interpreted differently by different people. Is a pipefitter someone who installs domestic plumbing or someone who lays large-bore main pipelines? An assessment should identify the actual work involved in the pipefitter's job.

Once the tasks required to accomplish a job have been determined, compare the requirements with the knowledge, skills and attitudes of the available personnel. You can then identify any gaps between the attributes required and the competence of existing personnel. There are three possible outcomes of this assessment:

- If the gaps can be eliminated through training, the assessment results can be used as the basis for designing a training programme.

Training

- If the gaps are great, then it may be necessary to recruit new, trained and competent staff.
- In some cases, where neither of the above options is feasible, it may be necessary to re-assess the broader strategy — for example, if the technology chosen is appropriate to local circumstances.

Time, funds and expertise may restrict the depth and detail of analysis. It is unusual to 'get it absolutely right' first time. So take a staged approach. Carry out a needs assessment, do some training, evaluate its effectiveness, assess needs again and further develop the training.

Training design

The training-needs assessment is used to establish aims and objectives based on the identified gaps in knowledge, skills and attitudes.

- An aim is an overall goal or statement of intent, for example, to increase the effectiveness of community water managers.
- General objectives describe what someone should know, understand, or appreciate at the end of a training session.
- Specific objectives clearly identify what participants will be able to do as an outcome of a training activity. For example, 'at the end of the session, participants will be able to assemble correctly a centrifugal pump'. Specific objectives are used to assess participants' performance: can they or can they not assemble a centrifugal pump?

A training session is designed around a general objective and a series of specific objectives.

Training methods

> I hear, I forget
> I see, I remember
> I do, I understand
>
> *Chinese proverb*

Participatory methods of training — in which people communicate with each other and learn by doing — are likely to be much more effective than one-way lecturing. Here is a selection of participatory training methods:

- *Buzz groups* — discussion in small groups on a particular topic, allowing everyone to be involved. Findings can be reported back to the whole group.
- *Brainstorming* — quickly generates ideas and responses which can be discussed after the brainstorm.
- *Case study* — trainer's presentation of an event or situation which participants discuss afterwards.
- *Role play* — participants act out a real-life situation, sometimes taking on an unfamiliar role.
- *Simulation* — a combination of case study and role play in which participants take on roles within a given scenario. Participants learn through their experience and analysis of the situation. De-briefing is an important part of the process.
- *Demonstration and practice* — participants observe a practical demonstration and then practise under close supervision.

Where and when should training take place?

Venue and timing can be crucial. Training in the workplace has its advantages:

- no trainee concerns about the expense, effort and time in travelling
- training specific to the circumstances of the trainees
- performance can be assessed in the work situation

But training in the workplace can also be distracting. A venue remote from the work environment allows participants to focus fully on the training, and share experiences with people from other organizations. Sharing experience, and establishing contacts and networks is an important aspect of training.

The importance of timing will vary but it must suit the participants if they are to attend and concentrate.

Who will do the training?

Competent individuals are often recruited from within an organization or community and trained as trainers. External agencies with experienced trainers may be required to assist in the development of training skills.

Training delivery

Training is 'delivered' through training sessions, courses and programmes. The manner of delivery can determine effectiveness. Trainees need to be relaxed and ready to participate. This means avoiding a top-down 'expert' approach. A trainer is a facilitator of learning, providing opportunities for participants to learn through experience, and to grow in confidence.

People are individuals, and learn individually, even in a group. To support each individual, a trainer needs to establish a rapport by:

- having a genuine interest in each trainee
- encouraging and enthusing
- involving everyone (trainers and trainees)
- ensuring that s/he can be clearly understood by everyone

Training

Training-course evaluation

The trainees, trainers, and managers all want to know if the training has been effective:

- were the objectives of filling the gaps in knowledge, skills and attitudes achieved?
- were participants satisfied with the training?
- was the training cost-effective?

There are several ways of assessing effectiveness, and a combination of methods may be used:

- questionnaires
- trainee presentations
- practical demonstrations

Impact evaluation

The final test of the effectiveness of training is whether what has been learned is applied in practice. To evaluate impact:

- assess work performance before training
- assess work performance after training

Information collected at the training-needs assessment stage, before training, can be useful at the evaluation stage.

The training process is not static. It should be one of continuous development. Situations change, and training itself is an agent of change. Evaluation results feed into the design of future training and so the training cycle continues.

On-the-job training

On-the-job training often relies on the assumed abilities of those involved in the hope that learning will take place, but with little thought given to how this should happen. In this situation the trainee is as likely to learn bad as well as good habits; the training must be planned.

Coaching is one method of planned on-the-job training. The trained coach (supervisor or counterpart) sets tasks and assignments, monitors progress, assesses performance and gives feedback. This is done within a planned framework.

Further reading

Bekkering, W., *Training and Teaching: Learn how to do it*, TOOL, Amsterdam, 1992.
Hope, A. and Timmel, S., *Training for Transformation: A handbook for community workers*, Mambo Press, Gweru, 1996.
Leigh, D., *Designing and Delivering Training for Groups*, Second edition, Kogan Page, London, 1996.
Pretty, J.N., Guijt, I., Thompson, J. and Scoones, I., *Participatory Learning and Action: A trainer's guide,* IIED, London, 1995.

Prepared by Jan Davis, Training Co-ordinator, RedR, 1 Great George Street, London SW1P 3AA, UK.
Phone:+ 44 171 233 3116   Fax: + 44 171 222 0564   E-mail: Bobby@redr.demon.co.uk

RedR relieves suffering in disasters by selecting, training, and providing competent and effective personnel to humanitarian agencies world-wide.

Supported by WEDC

WEDC Loughborough University Leicestershire LE11 3TU UK
www.lboro.ac.uk/departments/cv/wedc/   wedc@lboro.ac.uk

54. Emptying pit latrines

What to do with a full latrine pit

WARNING! When a single pit is full to within half a metre of the top, either:

■ stop using the latrine   or ■ empty the full pit

■ Back-fill the top of the pit with soil.

■ Dig another pit and build a new latrine.

by hand

- Dig out the contents using a spade and a bucket.

- Remember that if the latrine was used very recently, the excreta will be fresh and dangerous because it contains lots of pathogens (germs).

- By digging, you can become infected with worms and diarrhoea.

- Flies attracted to excavated material often carry infection to people nearby.

or by machine

- Use a big tanker with a powerful vacuum pump...

  Advantages
  - Your pit will be empty after only one or two tanker visits
  - The vacuum is strong enough to lift sludge from a depth of 2 to 3 metres

  Disadvantages
  - Large vehicles cannot negotiate narrow, twisting roads and alleys.
  - Vehicles with powerful pumps are very expensive, and it is often very difficult to get spares.

- ... or use a tanker with a less powerful pump which may be mounted on a Land Rover or similar vehicle ...

- ...or fill 200-litre drums with hand-operated pumps. Small teams of individuals may provide a private service to householders.

If a pit is to be emptied, it is usually 'lined' with walls of stones, bricks or concrete. If the pit is not lined, there is a danger of collapse when solids are removed.

If the sludge is too firm, jet on water and agitate the mixture of sludge and water with the end of the suction hose.

In Zimbabwe, for example, many unlined pits have collapsed when emptied.

Emptying pit latrines

Full twin-pit latrines

Make your single pit as big as possible,
- then it can be used for many years before filling up;
- you will have fewer problems with flies and smells; and
- the further down the excreta, the smaller the risk of disease.

> Many latrine pits in East Africa are more than 10m deep. Sometimes, pits 15 or 20m deep are dug in firm soil.

But twin-pit latrines can be fairly shallow — typically 1.5m deep. Each pit will take only two to three years to fill up, so use one pit until almost full — then use the other pit for the next two to three years.

Design and operation features of twin-pit latrines

1. Use completely separate twin-pit latrines alternately with pour-flush latrines.

2. Pour-flush latrines are built where people use water for anal cleaning. They take one or two litres of water to the latrine; a little is used for anal cleaning and the remainder is poured into the pan to flush faeces into the pit.

3. Excreta is flushed from the pan to a Y-junction and then to one of the twin pits.

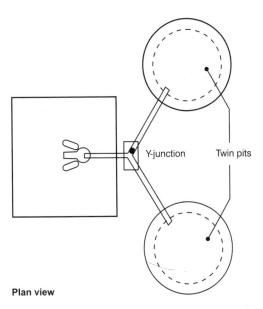

Plan view

4. When the second pit is full, the first pit is emptied. Because the content of the first pit has been maturing while the second pit fills, the excreta has become innocuous — it has lost any unpleasant smell and all pathogens have become inactive. As the pit is small and the sludge is no longer unpleasant or a health risk, the contents can be dug out by the family and neighbours.

5. When solid material such as leaves, grass or paper is used for anal cleaning, double or divided pits are suitable.

6. Walls extending for the full depth divide the pit into two or more sections. Each section is a separate chamber and is used for two or three years in the same way as one of the twin-pits.

7. For public or communal latrines, for example in schools, a long pit can be divided into several chambers. These chambers are used alternately for two or three years. Each chamber (except those at both ends) receives excreta from two cubicles.

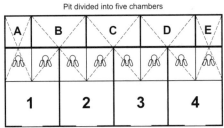

Pit divided into five chambers

Four cubicles

86

Emptying pit latrines

What to do with sludge from latrine pits and septic tanks

Faecal sludge varies in strength; this determines how you should treat it. High-strength sludge from unsewered public toilets or from bucket latrines commonly has a COD (Chemical Oxygen Demand) exceeding 20 000 mg/l, total solids greater than 3.5mg/l, up to 60 000 helminth eggs per litre, and NH_4-N ammonia concentration exceeding 2000mg/l. The ammonia concentration is particularly significant because liquid with a high ammonia content is likely to be toxic to the algae which are necessary for the operation of waste-stabilization ponds.

Common methods of dealing with sludge include:

Disposal into water
Sludge is regularly and indiscriminately dumped into rivers, ponds, lakes and the sea. *This is bad for the environment and a real health risk.*

Disposal onto land
Indiscriminate dumping onto land is as common and as undesirable as disposal into water. Untreated sludge can be used as a fertilizer *provided great care is taken to avoid contamination of crops.*

Composting
Mix sludge with two or three times its volume of vegetable waste. To keep it aerobic, turn it several times in the first few weeks. Then pile it into windrows (long heaps, often about 2m wide at the top, 2m high, with sides sloping at about 45°) for several weeks. You can then use it as a land-conditioner and fertilizer.

Household biogas units
Add latrine or septic-tank sludge to biogas units, whose main input is animal waste (e.g. cows in India or pigs in China).

Drying beds
Sludge flows onto a shallow tank to a depth of about 300mm. The base of the tank slopes to allow drainage and is covered with a layer of sand which forms a 'bed'. The time it takes for the solids content to increase until the sludge can be lifted by hand or mechanical shovel depends on temperature, humidity, and rainfall. In favourable conditions, this should be about a week.

If you have a wastewater treatment system you can also use:

Solids-liquid separation
A preliminary treatment. Dewater, dry or treat the separated solids by anaerobic digestion (as described below). Treat the liquid in waste-stabilization ponds, or mix it with municipal wastewater for conventional treatment.

Anaerobic digestion
Add sludge from pit latrines, aqua-privies and septic tanks to wastewater sludge separated by sedimentation at wastewater-treatment plants.

Extended aeration of septic-tank sludge (septage)
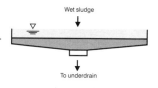
This method is good as less land is required for separation and dewatering. The capital and operating costs of aeration are high, but the smaller area involved should enable you to use local treatment units, and save money on haulage.

To sewerage systems
Sludge is often illegally dropped down manholes, increasing the likelihood of downstream blockage. Discharge stations can be specially constructed to receive and retain sludge from tankers, then discharge it to the sewer when the flow is appropriate.

Waste-stabilization pond systems
Faecal sludge may be treated in facultative waste-stabilization ponds together with municipal wastewater, or separately. The facultative ponds may be preceded by settling, thickening, and anaerobic ponds, and may be followed by maturation ponds.

Emptying pit latrines

Sludge from septic tanks

As sewage passes through a septic tank, heavy solids fall to the bottom, where a layer of sludge builds up. Light solids, like grease, rise to the surface and form a layer of scum.

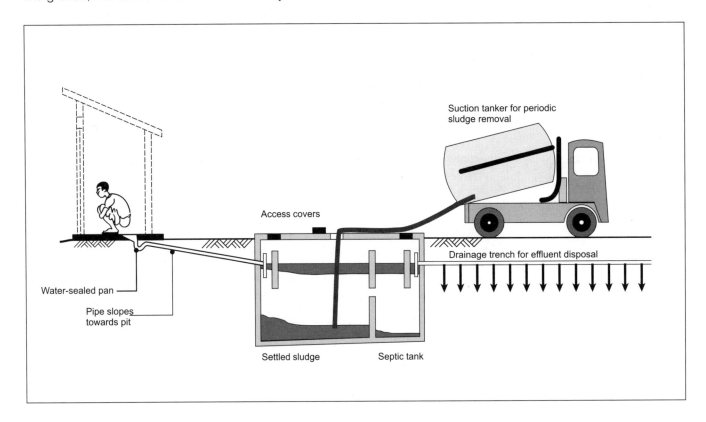

Sludge accumulates at the bottom of the tank. While lying there, it slowly decomposes by anaerobic digestion.

In tropical climates, decomposition may be rapid, and the net increase of sludge (accumulation less decomposition) slower.

After a certain time, the sludge must be removed — the tank is 'desludged'.

In temperate, industrialized countries, desludging is often carried out at regular intervals, for example, every six months.

In hot climates, you can leave the sludge for several years. A simple rule is to desludge the tank when the sludge occupies two-thirds of the tank volume.

Vacuum tankers are commonly employed for desludging septic tanks.

In general, septic-tank sludge is much less dense than solids removed from pit latrines. Consequently, a medium-powered vacuum pump is strong enough to lift septic-tank sludge.

Further reading

Franceys, R., Pickford, J. and Reed, R., *A Guide to the Development of On-site Sanitation,* WHO, Geneva, 1992.
Pickford, John., *Low-cost Sanitation: A survey of practical experience,* IT Publications, London, 1995.

Prepared by John Pickford and Rod Shaw

WEDC Loughborough University Leicestershire LE11 3TU UK
www.lboro.ac.uk/departments/cv/wedc/   wedc@lboro.ac.uk

55. Water source selection

Water is essential for life, but for many people, the quantity of water available may be minimal, and the water may be of poor quality. This Technical Brief outlines some of the issues which need to be considered when planning improvements to supplies, to ensure that the most appropriate sources of water are selected.

There are three types of water source: ■ **rainwater** ■ **surface water** and ■ **groundwater**

Rainwater

Collecting rainwater from either an existing roof structure or a ground catchment area can provide a useful supplementary source of water even if it is not used as the main supply. Storage tanks are usually required to make the best use of rainwater.

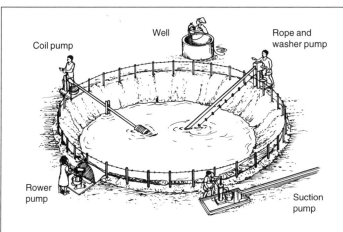

Surface water

When rain falls to the ground it becomes *surface water*, where it may move across the ground in the form of streams or rivers, or remain in one place in the form of ponds or lakes. Surface water is easily polluted and can be affected by wide seasonal variations in *turbidity* ('muddiness') and flow. Variations in turbidity present a challenge for the effective operation of treatment processes, while variations in flow affect the location and design of abstraction structures. Surface water, however, is often the easiest to access (see illustration).

Groundwater

Some surface water sinks into the ground and becomes *groundwater*. Here it can remain for a long time in an *aquifer* — spaces underground which can hold water because the surrounding earth and rock is *impervious* (does not let water through).

Groundwater may be obtained in several ways:

Water from mountain springs can often be transmitted to areas of demand by gravity, limiting the operation and maintenance requirements of a supply system.

Shallow wells can also provide a supply system with minimal operation and maintenance requirements — particularly if they are well-constructed, protected, and fitted with a handpump. For larger supplies, diesel or petrol pumps may be used in place of handpumps. Shallow wells can often be constructed using local techniques and labour.

Shallow or deep boreholes usually require drilling equipment and an experienced drilling team, but they can provide high-yield supplies of good-quality water. Groundwater, however, may be affected by high levels of chemicals, such as fluoride or chloride.

Locating groundwater can be difficult. The presence of existing wells with good, stable yields, other positive hydrological features, or information from satellite images can highlight groundwater potential but, following this, extensive field-trials are usually required to determine acceptable borehole locations.

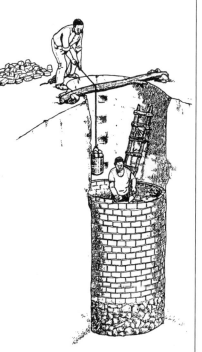

Lining a hand-dug shallow well

Water source selection

Selecting a water source for a community supply system requires careful consider[ation...] the existing sources in use are the most appropriate and only require minor impr[ovements...]

Socio-political and cultural considerations

Socio-political and cultural considerations are as important, if not more so, than the technical requirements for development. If the water supply is not culturally appropriate, and causes security difficulties or restricts access for certain groups such as women or disabled people, the benefits of the new system will be limited.

Women and water

Communities may use a single source or several sources of water for different needs such as drinking, washing clothes and watering crops. It is often the women and children who are most involved in water collection and its use. They are likely to have the most knowledge about existing sources, and are the people most likely to benefit if new supplies are developed. They are also the most likely to suffer if a new water-supply system is not appropriate to the needs of a community. It is essential, therefore, that women and children, as well as men, should be involved in every stage of a water-supply project.

Participative approaches should be used when selecting water sources and designing village-level supply systems. If some sections of the community are not involved and their views are not taken into account, the water-supply system is likely to be under-used and may easily fall into disrepair. People may revert to their old water sources which may be more polluted.

Water committees

Water committees are set up in many areas to manage water-supply systems. Care must be taken to ensure that all groups in the community are represented and can make their concerns and needs heard and understood. It is often difficult to achieve this. Women, for example, may form part of a water committee but they still may not have a voice within it because of cultural or social conditions which prevent them from speaking in public. Innovative approaches are required to ensure that representatives of as many groups as possible can participate equally.

Operation and maintenance

Care must be taken when identifying personnel both to undertake training, and to be responsible for operation and maintenance.

It is well documented that women often make the most conscientious maintenance workers but are often expected to undertake the task free of charge in situations where men would normally be paid. Care must be taken, therefore, to ensure that both women and men are consulted on the matter, that they are willing to undertake the task, and that they are compensated in a way which is fair and appropriate.

Yield versus demand

The yield must be adequate. If a more convenient supply is developed, then consideration must be given to the potential increase in demand and to the possible migration of outsiders into the community, particularly in areas where water is scarce.

Socio-political and cultural considerations

○ Has a thorough assessment been undertaken of the needs and wishes of the community, involving all groups (women, men and children and members of any distinct social groups, particularly those who are most vulnerable due to their gender, caste or class)?

○ In the village, who does what, where and when?

○ Who controls, and who owns resources?

○ What are the power structures within the village, and how will they impact on the use and benefits to be gained from the development of the source?

○ Are there barriers to the involvement of any groups in the assessment, design, construction, operation and maintenance, and evaluation of any system?

○ Is the planned system culturally acceptable to all groups?

Water quality

○ What is the existing, seasonal and predicted future water quality?

○ How easily can the source be protected against pollution?

○ What is the required quality?

○ What treatment is required and is it feasible in the village context?

Impacts of development on:

○ the health of women, men and children?

○ the economic status of women, men and children?

○ time available to women, men and children?

○ the environment, e.g. on the aquifer or on vegetation and erosion?

○ domestic and wild animals?

Yield

○ Does it have an [adequate] present deman[d]

○ Does the yield [...]

○ Is the yield expe[cted...]

○ Is the demand e[...]

Consi[derations for] selectin[g a source] for a water-[supply...]

Water source selection

of a range of factors. The illustration below highlights some of these. It may be that
nt. In other cases, a new source or sources may have to be developed.

Water quality

All water is susceptible to contamination. It may accumulate contaminants from the air, the ground, or from rocks. Some of these contaminants, such as low levels of certain minerals or compounds, are not harmful to health, whereas others, such as pathogens, may be.

The water quality must also be acceptable and treatment methods suited to the community concerned. What local treatment methods, if any, are already being used in the area? Can they be used in the new system? The benefits of using improved sources of water will be increased if the community practises good sanitation and hygiene. Will their current behaviour pollute the water source or reduce the benefits of an improved supply? Would additional resources be required to help reduce these risks? Some water-quality problems such as high fluoride levels are very hard to treat and have serious health implications, whereas others, such as turbidity, are usually easier to deal with.

Technical requirements

The development of the source must be technically feasible, and the operation and maintenance requirements for the source abstraction and supply system must be appropriate to the resources available. If the supply system cannot be operated and maintained either by the villagers themselves or the organizations or institutions within the area, then the systems are likely to be misused or fall into disrepair.

Economic considerations

Care must be taken to ensure that funds are available for both the construction and the operation and maintenance of the system over the longer term. Who will pay, how will they pay, and how much will they pay? Who will manage and maintain the system, and who will collect the funds? From whom will the resources be obtained and how will they be secured?

Legal and management requirements

Current ownership of the land and the legal requirements of obtaining permission to abstract are also factors to consider when selecting a source. Sources on private land may cause access problems for certain groups which may not be apparent at the outset. The consequences of siting decisions must be considered carefully.

Impacts of development

The use of a particular water source will have impacts on the people who use it, on animals, and on the environment. The impacts on people may be positive or negative, and may be related, amongst other things, to health, economic status or time. If a surface-water source is used, there may be impacts on remote users and, likewise, if wastewater enters surface-water sources, there may be similar impacts. Impacts on the environment may include loss of vegetation, erosion, or the draining of an aquifer.

Technical requirements for development and for operation and maintenance

- Have the users been involved in the planning and design of the system?
- Details for:
  - protection
  - abstraction
  - treatment
  - transmission
  - storage
  - distribution
  - subsidiary requirements?
- Are the resources (both human, equipment and material) available?
- Are the techniques already used locally? Who will be involved in the construction and operation and maintenance of the system (women, men and /or children)?
- Is the required training available?
- Can the system be constructed locally, or will outside support be required for construction and for operation and maintenance over the long term?
- Will the supply be accessible for all members of the community, especially for the main users of water and those who may have accessibility problems such as the aged or disabled members of the community?

Economic considerations

- What will be the financial cost of the system (both capital, and operation and maintenance)?
- Who will pay (individuals or organizations within the community or outside organizations)?
- How much are they willing to pay?
- Who will, potentially, benefit economically from the new system?
- Who will, potentially, lose economically from the new system?

Legal and management requirements

- Who owns the land?
- What are the legal requirements to obtain permission to abstract?
- What are the management requirements for the system?
- Who will manage it?
- Will they require additional training and support?

Water source selection

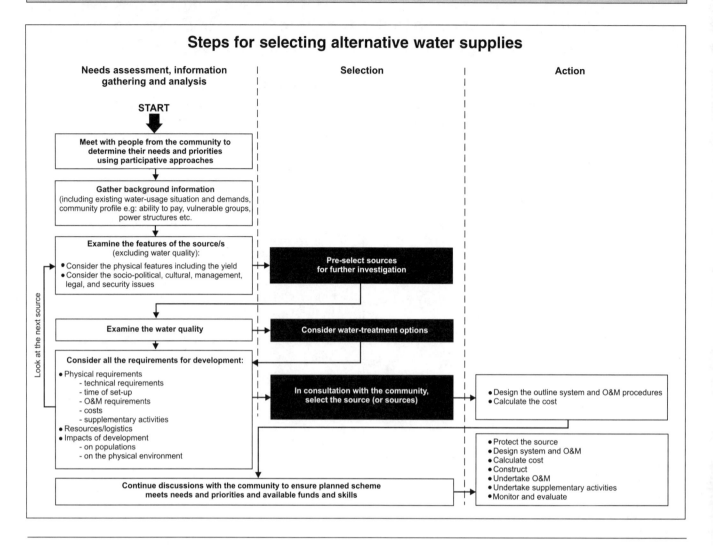

Further reading

Almedom, A. and Odhiambo, C., 'The rationality factor: Choosing water sources according to water uses', *Waterlines*, Vol.13, No.2, IT Publications, London, 1994.

Brikké, F., Bredero, M., de Veer, T., and Smet, J., *Linking Technology Choice with Operation and Maintenance for Low-Cost Water Supply and Sanitation*, IRC International Water and Sanitation Centre / World Health Organization / Water Supply and Sanitation Collaborative Council, Geneva, 1997.

Cairncross, S., and Feachem, R., *Small Water Supplies*, Bulletin No. 10, Ross Institute, London, 1978.

House, S.J. and Reed, R.A., *Emergency Water Sources: Guidelines for selection and treatment*, Water, Engineering and Development Centre (WEDC), Loughborough, 1997.

IRC (1994a) *Together for Water and Sanitation: Tools to apply a gender approach,* Occasional Paper Series 24, IRC International Water and Sanitation Centre, The Hague, 1994.

IRC (1994b) *Working with Women and Men on Water and Sanitation: An African field-guide*, Occasional Paper Series 25, IRC International Water and Sanitation Centre, The Hague, 1994.

ITDG Water Panel, 'Guidelines on the planning and management of rural water in developing countries', *Waterlines,* Vol.7, No.3, IT Publications, London, 1980.

Prepared by Sarah House, Bob Reed and Rod Shaw

WATER AND ENVIRONMENTAL HEALTH AT LONDON AND LOUGHBOROUGH (WELL) is a resource centre funded by the United Kingdom's Department for International Development (DFID) to promote environmental health and well-being in developing and transitional countries. It is managed by the London School of Hygiene & Tropical Medicine (LSHTM) and the Water, Engineering and Development Centre (WEDC), Loughborough University.

Phone: +44 1509 222885   Fax: +44 1509 211079   E-mail: WEDC@lboro.ac.uk   http://www.lboro.ac.uk/well/

56. Buried and semi-submerged water tanks

This Technical Brief outlines the advantages and disadvantages of using buried and semi-submerged tanks for collecting and storing water. It also examines some of the design features and construction procedures.

Advantages of an underground tank

- If the soil is firm it will support the pressure acting on the walls of the tank so that cheaper walls — less robust than those of an above-ground tank — can be used;
- the tank is protected from the cracking which can result from the regular expansion and contraction caused by daily heating and cooling of exposed walls;
- the water in it remains cooler and is, therefore, more pleasant to drink; and
- water can be collected from ground-level catchment areas.

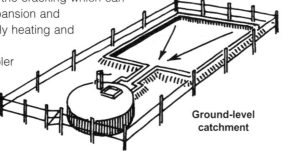

Ground-level catchment

Disadvantages of an underground tank

- The source of any leakage is hard to detect and, therefore, hard to repair;
- polluted water may leak into the tank, particularly if the roof is buried;
- drawing water from a tap (more hygienic than using a bucket and rope) is only possible if steps are provided to give access to a low-level tap in a trench immediately adjacent to the tank (see right). If the buried tank is on a hillside, however, water will gravitate to an above-ground tap (see below right);
- if the level of water in the ground around the tank ever reaches a high level, an empty tank could float out of the ground like a boat!

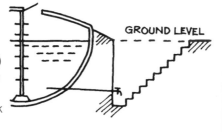

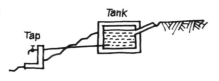

Construction materials

Materials in common use include:
- clay or thin, impermeable, man-made membranes used to line excavations;
- brickwork, blockwork or stone masonry — particularly for walls, and occasionally for arched roofs;
- reinforced concrete for walls, and floor and roof slabs; and
- ferrocement for walls, roofs and, sometimes, floors.

Other matters to consider

- two tanks, or a tank divided into two compartments, allow one tank to be maintained while the other continues to provide water;
- access manholes should have covers which can be locked and prevent contamination;
- if the floor of a tank slopes to a low point at which a pipe outlet is provided, it is easy to wash out any sediment that may collect in the tank.

Roofs

- Prevent nearly all evaporation;
- protect potable water from contamination and algae growth;
- prevent the breeding of mosquitoes — but only if *all* openings to the air are screened with mosquito mesh;
- can be exposed or buried (buried roofs must be very strong to withstand the weight of a vehicle);
- flat roofs are often made from reinforced-concrete (RC) slabs. Larger spans need RC beams and column supports;
- thin, domed ferrocement roofs are usually more cost-effective than flat roofs — they utilize the high compressive strength of the cement mortar. The mortar is reinforced with welded and woven wire meshes; and
- lightweight materials such as corrugated iron can also be used for exposed roofs, but timber supports are not recommended as they are liable to rot.

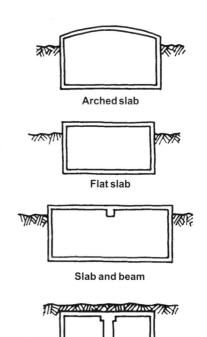

Buried and semi-submerged water tanks

Outline construction notes for a partly submerged, hemispherical ferrocem

Important notes

Ferrocement is a mixture of cement mortar and wires. This design is a particularly cost-effective method of in-ground tank construction. Full construction details are found in Nissen-Petersen (1992). See the other books listed on page 96 for more detailed advice on producing good concrete and good ferrocement.

- Select clean sand carefully — it must not be too fine;
- keep ferrocement damp between the application of different layers and for some time (ideally, three weeks) after applying the last layer. This 'curing' is also important for concrete. Where possible, use polythene sheeting (or wet sand on the floor and roof) to reduce evaporation of curing water;
- for the ring-beam, use a 1:3:4 concrete mix (i.e. 1 volume measure of cement: 3 measures of coarse sand: 4 measures of stones graded up to 25mm);
- use a mortar mix for the ferrocement of 1:3 (i.e. 1 volume measure of cement : 3 measures of sand). Measure volumes carefully, and keep the water content as low as possible;
- apply 'nil' (a mixture of water and cement with a porridge-like consistency) to improve the water-tightness of the ferrocement.

Main materials needed

- Cement: 73 bags
- BRC welded mesh No.65 (5.4mm diameter bars on a 150mm grid): 2m x 35m
- 50mm GI pipe for roof support: 4.5m
- oil drums for sheets: 48
- barbed wire: 1.6mm wire, 25kg
- chicken mesh (25mm holes): 0.9m x 175m
- access cover: 1
- handpump or pipework and tap for water collection
- timbers; poles; flat irons; angle irons and 'u' bolts for ladder on king-post
- stones for wall, aggregate for concrete, sand for concrete and ferrocement
- binding wire: 2kg
- polythene: 2m x 30m
- nails 50mm: 2kg
- sand: 17 tonnes
- concrete aggregate: 1 tonne
- stones for wall: 12 tonnes

1
- Mark out the circumference of excavation on the ground using a 3.12m length of string on a peg at least 10m away from any trees.

2
- Excavate 3.12m radius hemisphere around a temporary pillar of soil which remains to support the peg.
- The soil must be firm to support the tank.

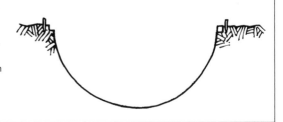

3
- Offset pegs 250mm from the edge of the excavation.
- Mark horizontal line on pegs.
- Excavate a 200mm x 200mm shelf for wall foundation.

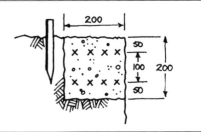

4
- Cast a level-concrete (1:3:4 mix) ring-beam foundation with two layers of four strands of barbed wire.
- Cure.

5
- Build a 0.6m-high horizontal wall on the ring-beam, using bricks or stone masonry.
- When it is firm, wrap 12 strands of barbed wire round tightly and cover with cement mortar (1:3 mix).
- When the mortar has hardened, cure and then pile the excavated soil against the wall.

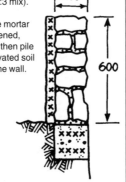

6
- Apply 10mm thickness of cement mortar to the soil and to the inside face of the wall. Cure.
- A day later, add a second 20mm-thick layer of mortar. Cure.

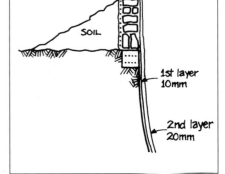

Buried and semi-submerged water tanks

Water tank of 80m³ capacity

7
- Next day, nail a spiral of barbed wire onto the mortar using 50mm nails. The spacing between the wires should be 200mm. Then add radial wires with a maximum spacing of 300mm at the top of the wall. These wires should project above the wall by 300mm (so they can later be incorporated into the roof).
- Now cover the inside of the tank with at least one layer of chicken mesh (with 25mm apertures) with overlaps of at least 200mm between adjacent layers. If using a tap, install the pipe into an adjacent excavation and construct steps to reach the tap.

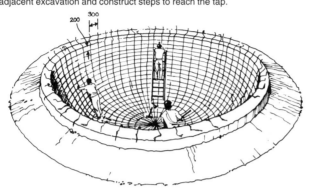

8
- Apply 30mm of 1:3 mortar to cover the inside of the hemisphere and, on the same day, add a thin layer of nil. Cure for at least three weeks — you can still work on the roof.
- The arched roof is supported by a 50mm-diameter galvanized-iron pipe which projects for 1m above the wall around the tank. Later, bolt angle-irons onto this 'king post' to act as a ladder below the access cover, near the centre of the roof. Position the pipe at the centre of the tank on top of two, crossed, flat irons which spread the load onto the floor; then cover these irons with mortar.
- Fix two more crossed flat irons across the top of the pipe which projects 25mm above the roof shuttering so that the irons are cast into the roof.

9
- For roof shuttering (formwork) erect a minimum of 12 radial timbers **(a)**, 150mm x 25mm shaped to the curvature of the roof (a vertical radius of 6.25m), strengthened with a 50mm x 100mm timber **(b)** on each side, to span from the wall to vertical timbers **(c)** tied to the king-post.
- Strengthen the tapered ends of the radial timbers near to the walls with short pieces of 10mm-diameter bar **(d)** wired onto their underside.
- Support the ends near the wall with more vertical posts **(e)** 1.9m long.
- Nail some 50mm x 50mm timbers **(f)** on the 50mm x 100mm radial timbers to stiffen the formwork and support the edges of the steel-sheets **(g)** used for shuttering; these sheets can be cut from flattened 210-litre oil drums (you need 48), but ensure that one edge has a curve of 3.06m radius while the corners on the other edge are cut off as shown.

- Cut to size about 50 additional poles **(h)** (sisal poles are ideal) and use vertically inside the tank to support the sheets and make a strong curved formwork. Wire, but do not nail, so you can remove the shuttering more easily.

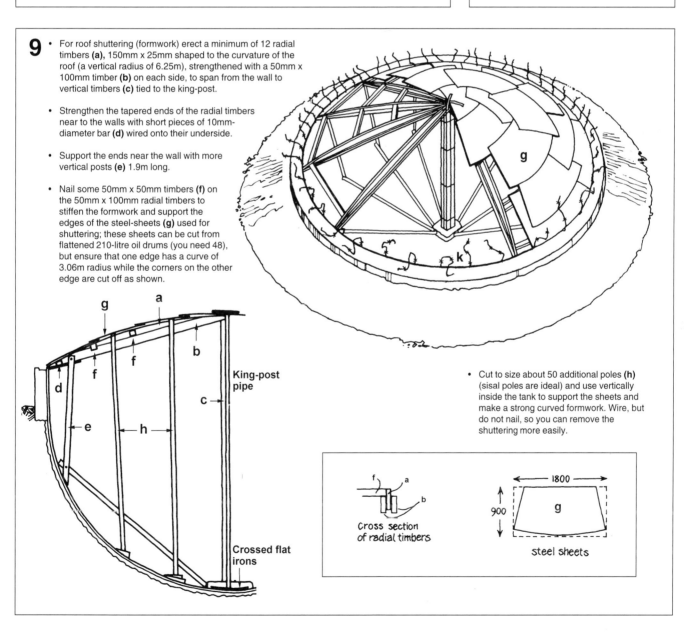

Buried and semi-submerged water tanks

10
- Cover the dome with polythene, followed by trapezium-shaped pieces of welded mesh **(j)**. The mesh is BRC mesh 65 (i.e. 5.4mm diameter bars at 150mm centres).

- Wire adjacent sheets of mesh together with an overlap of at least 200mm. The barbed wire **(k)** from the walls (see Box 9) is tied to the welded mesh and everything is covered with one layer of chicken mesh **(l)** with 200mm overlaps. The whole roof is now covered with a 50mm layer of well-compacted cement mortar. The reinforcement is lifted into the centre of this layer before compaction is completed. A curved timber **(m)** is rotated around the centre of the roof to get the right shape.

- Cure the roof for at least three weeks, although you can remove the supports after ten days. Once the shuttering and polythene is removed, apply mortar as necessary to any patches under the roof which need repairing so that all reinforcement is properly covered. Seal the joint between the wall and the dome with cement mortar.

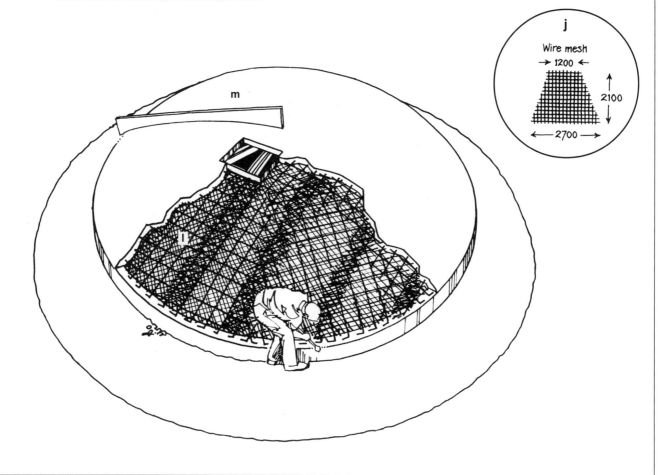

Further reading

Reed, R.A., Shaw, R.J. and Skinner, B.H., Ferrocement water tanks, Technical Brief No. 36, *Waterlines,* Vol.11 No.4, IT Publications, London, 1993.
Watt, S.B., *Ferrocement Water Tanks and their Construction*, IT Publications, London, 1978.
Nissen-Petersen, E., *How to build an underground tank with dome*, ASAL Consultants Ltd., PO Box 38, Kibwezi, Kenya, 1992.

Prepared by Brian Skinner and Rod Shaw

WATER AND ENVIRONMENTAL HEALTH AT LONDON AND LOUGHBOROUGH (WELL) is a resource centre funded by the United Kingdom's Department for International Development (DFID) to promote environmental health and well-being in developing and transitional countries. It is managed by the London School of Hygiene & Tropical Medicine (LSHTM) and the Water, Engineering and Development Centre (WEDC), Loughborough University.

Phone: +44 1509 222885   Fax: +44 1509 211079   E-mail: WEDC@lboro.ac.uk   http://www.lboro.ac.uk/well/

57. Surface water drainage — How evaluation can improve performance

This Technical Brief outlines ways in which, by carrying out a simple evaluation, engineers and technicians can make improvements in the performance of drainage systems.

Urban drainage is the removal of unwanted water from cities and large towns. When it rains, part of the rainwater, called *runoff*, runs off the surface and flows along the ground. Surface-water drainage removes this runoff.

Without surface-water drainage, frequent flooding creates many problems:

- floods damage roads, houses, and goods at major cost;
- during floods, runoff mixes with the human wastes inside latrines, septic tanks and sewers, and spreads them wherever the runoff flows; and
- mosquitoes breed in ponds (even small ones!) that are not drained within a week, so contributing to malaria and other diseases.

Flooding can occur where drains are:

- poorly designed;
- poorly built; or
- blocked with solids such as rubbish, or broken brick, bits of concrete, soil, and human wastes.

NO drainage system can protect residents from all storms. In many cases, however, drainage does not work as well as it could, so there is unnecessary flooding.

Why evaluate drainage?

Evaluations can answer such questions as:

- Is flooding a problem in this area?
- Are drains blocked? With what?
- How does the drainage system work in practice?
- Is maintenance a problem? Can it, realistically, be improved?

Drainage evaluation methods

Is flooding a problem in this catchment?

There are two useful approaches: asking residents (resident surveys), and seeing for yourself (direct observation).

Resident surveys

People who have lived in one area for several years know a lot about flooding – they remember when water flooded their homes. You can get an idea of which areas are worst affected by simply talking to people. There are simple rules:

Avoid 'leading' questions

People's answers reflect what and how they are asked. Questions must be open and neutral, allowing each person to express him or herself freely; 'What happens when it rains?' is better than 'Does it flood a lot here?'

Ask more than one person

If just one or two people are asked, they will know some parts of the area better than others. If men work outside the area, and women spend more time in the home, women will know more about minor flooding.

Try to be specific

It is probably better to ask first about last year's flooding, rather than 'how high does water rise?' It is also best if residents find a specific place to show the high water mark, rather than stating that 'water was knee-deep.'

Direct observation

Walking around in a storm can be a good way to see what happens when it floods. It is a fairly limited exercise, however, because:

- you can only do in the rain;
- you cannot be everywhere all the time; and
- you can easily miss the most important part of the storm.

Direct observation during floods is more helpful in getting a feel for how the system works as a whole, than for gauging severity accurately.

Surface water drainage

Are the drains blocked?
The best way of finding this out depends on the type of drain — open drains are much easier to check than closed ones.

Open drains
If open drains are used only for runoff, they are dry in dry weather. A quick walk along the drain can give you a good idea of the extent of the blockage. Frequently, however, open drains carry sewage as well as runoff. While a quick look can find a complete blockage, it cannot tell you much about the solids below the surface. A survey, using simple equipment to gauge the amount of blockage, can be helpful (see Figure 1 below).

In any drain where there are substantial solids, parts of the drain must be cleaned out to find the true depth to the bottom. Forcing a steel rod through deposits until you 'hit bottom' will *not* work, as the rod may lodge itself on top of a rock or brick, rather than at the true bottom of the channel.

Closed drains
Finding blockages in closed drains is more difficult, especially if they also carry sewage. Here are two quick checks:

Standing-water checks in manholes
When water is found standing in a manhole above the bottom ('invert') of the outgoing pipe, then something is holding up the flow (see Figure 2 on page 99).

Lamp-and-mirror checks
Where manholes are spaced less than 30m apart, lowering a powerful lamp down one manhole, and a mirror down another can be helpful (Figure 3). If the pipe is clear, the light can be seen clearly in the mirror; if the pipe is blocked with solids, or is not straight, then the light will be partly or completely blocked. Success depends on having a powerful lamp, which you must keep dry or the batteries will run down too quickly.

How does the drainage system behave in practice?

To get the clearest idea, look at how the system works in a storm.

Problem areas for flooding
Systematic observation is difficult unless problem areas have been identified before the storm. Define these using resident surveys *before*

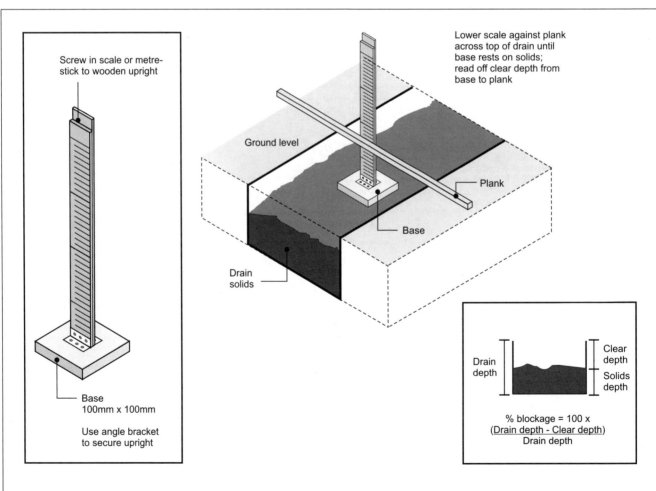

Figure 1. A scale for measuring the depth of solids in an open drain

% blockage = 100 × (Drain depth − Clear depth) / Drain depth

Surface water drainage

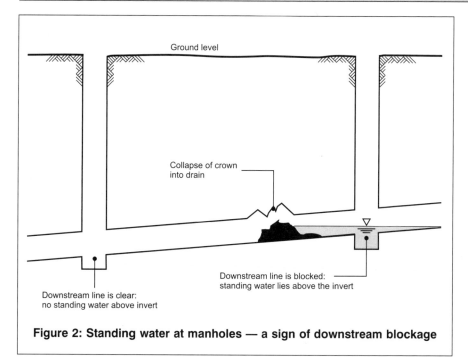

Figure 2: Standing water at manholes — a sign of downstream blockage

Figure 3: Use of a lamp and a mirror for checking drain clearance

The drainage 'network' is more than just the drain; it includes surface and gutter flow, inlet flow, and whatever is going on downstream, too.

The surface flow routes followed by runoff during floods

Runoff follows surface routes during floods. Study these routes during storms to find out both the impacts on residents, and ways to reduce problems. In some cases, flow leaves one drain and re-enters another with no problem; in other cases, whole areas become flooded.

Sometimes, small changes in such routes, for example, by raising a dyke or removing some soil, can improve the situation significantly. But someone must step back and look around to ensure that the problem of five houses is not being solved at the expense of 20 others!

Working in wet weather

Organizing a team to study drainage during storms

Storms are unscheduled, chaotic, and unpleasant; staff must be organized to work well in bad weather. The manager should assign tasks and responsibilities for the 'next storm' during dry weather – team members then know where they have to go and what they have to do at the start of the next storm *without* having to assemble as a group.

Checking catchment and sub-catchment boundaries

Good maps make this job much easier. Each team member should be allocated a 'beat', and should note on a map the direction of flow along the surface early in the storm. This should be completed within the first storm or two of the season, to define the catchment as early as possible.

the storm, and assign specific team members to observe them during rain. This can often clarify the cause of the flooding, such as inflows from other areas, or blocked inlets.

The hydraulic performance of the total drainage system

The only way to observe hydraulic performance is to study the drain itself during storms. Such a survey can find:

- overflow locations;
- bottlenecks and high head losses, eg culverts; and
- obstructed entry to the drain, (inlet blockage, poor inlet design, or poor surface grading).

Surface water drainage

Summary
Table 1 (right) shows how different data can be gathered at different stages of a storm. In practice, no team can count on a flood occurring, but its members can be ready when a flood takes place.

Improving performance from evaluations
Here are a few examples from experience in the Madhya Pradesh city of Indore:

Table 1. Wet weather observations and timing

	Beginning (or small storms)	Middle (flood)	End (flood-water drainage)
Catchment boundaries	Directions of flow in streets	Catchment boundaries	Catchment boundaries
Flood-prone areas		Bottlenecks, causes of flooding	
Hydraulic performance	Inlet performance, blockages, high head losses	Outflows from drains	Problems of grading, slow drainage, high head losses
Surface routes		Location, direction, and magnitude of surface routes	Location, direction and magnitude of surface routes
Nuisance and hazards		Observe, discuss with residents	Observe, discuss with residents

Understanding the catchment better
The designer may have missed some of the area that contributes runoff. Field evaluations can establish this quickly, and suitable diversion strategies can then be developed.

Solids-depth monitoring
The initial survey of solids depths can identify the first priorities for cleaning. Follow-up surveys can monitor how quickly solids build up after cleaning, and whether cleaning needs to be more frequent.

Blockages
A lamp-and-mirror survey can be a quick and efficient way to get an idea of the condition of old drains. One of the Indore surveys identified several problems within a few hours.

Surface routes of flow
Drainage designers usually focus on the routes of the pipes and channels, and not on the way water flows over the ground during a flood. Minor changes in some street levels can make a big difference to how quickly they drain after a storm.

Acknowledgements
These findings grew out of Research Project R5477, *Performance-Based Evaluation of Surface Water Drainage in Low-Income Communities,* of the Engineering Division of the Department for International Development (DFID). DFID, however, does not accept any responsibility for any of the information provided, or views expressed.

The author acknowledges the help of Dr David Butler, Mr Jon Parkinson, Professor T.A. Sihorwala and the staff of the Indore Drainage Evaluation Project in the development and testing of these approaches.

Further reading
Cairncross, S. and Ouano, E.A.R., *Surface Water Drainage for Low-Income Communities,* WHO, Geneva, 1991.
Cairncross, S. and Ouano, E.A.R., 'Surface water drainage in urban areas', in *The Poor Die Young,* edited by J.E. Hardoy, S. Cairncross, and D. Satterthwaite, Earthscan, London, 1990.
Cotton, A. P. and Tayler, K., *Urban Upgrading: Options and procedures for Pakistan,* WEDC, Loughborough, 1993.
Kolsky, P., *Storm Drainage: An engineering guide to the low-cost evaluation of system performance,* IT Publications, London, 1998.

Prepared by Pete Kolsky and Rod Shaw

WATER AND ENVIRONMENTAL HEALTH AT LONDON AND LOUGHBOROUGH (WELL) is a resource centre funded by the United Kingdom's Department for International Development (DFID) to promote environmental health and well-being in developing and transitional countries. It is managed by the London School of Hygiene & Tropical Medicine (LSHTM) and the Water, Engineering and Development Centre (WEDC), Loughborough University.

Phone: +44 1509 222885 Fax: +44 1509 211079 E-mail: WEDC@lboro.ac.uk http://www.lboro.ac.uk/well/

58. Household water treatment 1

This Technical Brief is the first of two examining the treatment of water in the home. Here we introduce the subject, and cover treatment by straining, storage, settlement, solar disinfection, chemical disinfection, and boiling. The second Brief (No. 59) considers treatment by coagulation, flocculation, filtration and solar distillation, and covers aspects of the reduction of some chemical concentrations.

Why treat water?
It is always better to protect and use a source of good quality water than to treat water from a contaminated source ...

but water needs to be treated if:

- people refuse to use it because of its colour or taste; and / or
- chemicals or organisms in it pose a health risk to users.

Check if a single communal system which treats water for everyone would be more cost-effective than a small system in every home. Remember that the community must be willing to co-operate in operating and maintaining a communal facility properly.

What contaminates water?
Pathogens (disease-causing organisms) including eggs or larvae of parasitic worms; bacteria; amoebas; and viruses.

Harmful chemicals from human activities (e.g. pesticides and fertilizers) or from natural sources (e.g. chemicals from rocks and soils).

Contaminants or physical properties which, although not harmful, cause people to reject the water because of its taste, smell, colour, or temperature.

What is the best source?
Where there is no good source the best option is to treat water from the source with the highest quality water. A change of source, or use of a treatment process, however, may give the water a different taste, unacceptable to the community.

Surface water is usually quite badly contaminated (see page 89).

Groundwater is usually much purer than surface water, but may be contaminated by natural chemicals, or as a result of human activities (including the unhygienic use of a bucket and rope in a well).

Rainwater captured from roofs made of sheets or tiles is relatively pure, particularly if the first water to flow off after a dry period is run to waste (see *The Worth of Water*, pages 45-48).

How much water needs to be treated?
Check first if it is feasible to only treat water used for drinking or preparing food which is eaten uncooked. Usually less than 5 litres/person/day are needed for these purposes. Providing only this amount of treated water will be much easier than treating all the water used in the house.

If the raw water looks reasonably clear it will not *usually* need to be treated before being used for other domestic purposes.

Water may sometimes need treatment for:

- **bathing** – if it contains pathogens which penetrate the skin (such as *cercariae*, which transmit schistosomiasis);

- **cooking** – if excessive iron or manganese cause problems with taste or colour, or if harmful chemicals are transferred from the water to the food;

- **laundry** – if it contains so much iron or manganese that it stains the clothes.

Maintaining the quality
Removing pathogens will be pointless if the treated water is contaminated again before it is drunk. The treated water should be carefully stored in, and hygienically collected from, covered containers.

101

Household water treatment 1

Straining

Pouring turbid (cloudy) water through a piece of fine, clean cotton cloth will often remove a certain amount of the suspended solids contained in the water. If the cloth is dirty, additional pollutants may be introduced! Purpose-made monofilament filter cloths can be used in areas where guinea-worm disease (dracunculiasis) is endemic. Such cloths are effective in straining out the copepods in the water. These tiny water creatures act as intermediate hosts for the larvae which transmit the disease. Some guinea-worm eradication projects supply a large-diameter drinking-straw with a filter mesh on one end so that copepods are strained out when water is sucked up the straw.

Vigorous shaking

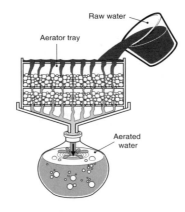

Aerator trays

Aeration

Aeration increases the air content of water; it reduces the concentration of volatile substances, such as hydrogen sulphide, which affect the odour or taste of water, and oxidizes iron and manganese prior to settlement or filtration. Dissolved air is also important for the effective performance of slow sand filters but there may already be sufficient oxygen in surface waters. You can achieve aeration on a small scale by rapidly shaking a vessel part-full of water. Aerate larger volumes of water by allowing them to trickle through one or more well-ventilated, perforated trays containing small stones.

Storage and settlement

Storing water for just one day can result in the die-off of more than 50 per cent of most bacteria; conditions in storage vessels are usually not conducive to their survival! Longer periods of storage will lead to further reduction. The *cercariae*, which are an intermediate host in the life cycle of schistosomiasis, can only live for 48 hours after leaving a snail if they do not reach a human or animal host. So storing water for more than two days effectively prevents the transmission of this disease.

The suspended solids and some of the pathogens in any water left in a container will begin to settle at the bottom. After several hours water collected from the top of the container will be relatively clear, unless the solids are very small (e.g. clay particles). Although this clear water will still contain some pathogens, many others will have settled to the bottom, often attached to the surface of particles. The three-pot treatment system exploits settlement and the death of pathogens during storage to improve the quality of raw water.

Disinfection

Disinfection is a way of ensuring that drinking-water is free from pathogens. For chemical and solar disinfection to be effective – and to a lesser extent for boiling – the water should be free of organic matter and suspended solids. Hence disinfection should be the *final* treatment stage, after any other treatment processes.

Disinfection by boiling

A typical recommendation for disinfecting water by boiling is to bring the water to a rolling boil for 10 minutes. As Miller (1986) points out, reaching 100°C for a few moments will kill most pathogens and most are killed at much lower temperatures (such as 70°C). The main disadvantage of boiling water is that it uses up fuel. It also affects the taste of the water, although increasing the air content by vigorously stirring the water, or shaking it in a bottle, after cooling, will improve the taste.

The Three-Pot Treatment System

Drinking-water: Always take from pot 3. This water has been stored for at least two days, and the quality has improved. Periodically this pot will be washed out and may be sterilized by scalding with boiling water.

Each day when new water is brought to the house:
(a) Slowly pour water stored in Pot 2 into Pot 3, wash out Pot 2.
(b) Slowly pour water stored in Pot 1 into Pot 2, wash out Pot 1.
(c) Pour water collected from the source (Bucket 4) into Pot 1. You may wish to strain it through a clean cloth.

Using a flexible pipe to siphon water from one pot to another disturbs the sediment less than pouring.

Household water treatment 1

Effectiveness of different household treatment systems

The effectiveness of some of the treatment systems shown in the table has been generalized. A treatment method should always be properly tested in a field situation before it is widely promoted. Some of the methods need to be combined to be effective.

Effectiveness of treatment method
0 = minimal if any effect. 1 - 4 = increasing effectiveness.
- = unknown effect. + = helpful to another process mentioned.

Problem with raw water	Straining through fine cloth	Aeration	Storage / pre-settlement	Coagulation, flocculation and settlement or filtration	Fine sand filtration (slow)	Coarse sand filtration (rapid)	Charcoal filter	Ceramic filter	Solar disinfection	Chemical disinfection	Boiling	Desalination / Evaporation
PATHOGENS												
Bacteria, (effectiveness often also apply for amoebas, viruses and ova)	0	+	1 - 2	0 - 1	4	2	-	3 - 4	4	4	4	4
Guinea-worm larvae (in cyclops)	4	0	0	-	4	2 - 3	-	4	2-4 b	-	4	4
Schistosomiasis cercaria	-	0	4	-	4	2 - 3	-	4	2-4 b	4	4	4
NATURAL CHEMICALS a												
Iron and manganese	0	+	1	1	3	3	-	-	-	-	-	4
Fluoride	0	-	0	4			-	-	-	-	-	4
Arsenic	0	+	-	4	4	4	-	-	+	-	-	4
Salts	0	0	0	-	0	0	0	0	0	0	0	4
OTHER PROBLEMS												
Odour and taste	0	2	1	1	2	2	3 - 4	2	0	1	-	3 - 4
Organic substances	1	1	2	1	3	3	-	3	-	4	-	4
Turbidity (cloudiness produced by suspended solids)	1	0	2	3	4	3	-	4	0	0	0	4

a. See pages 107 - 108.
b. Removal depends on sufficient temperature rise, cercaria die at 38°.

Chemical disinfection

Chlorination (see TB 46) is the most widely used method for disinfecting drinking-water. There are several different sources of chlorine for home use including liquids (such as bleach), powders (such as bleaching powder) and purpose-made tablets. Iodine, another excellent chemical disinfectant, is also used sometimes. With both of these chemicals you must add sufficient to the water to destroy all the pathogens; deciding on the right amount can be difficult since it will depend on substances in the water which will react with the disinfectant, and which may vary from season to season.

Another complication is that the strength of the chlorine compounds will vary with time; air-tightness, low temperature and absence of light are

Strength of various chlorine preparations

	Example calculation			
	% active chlorine when fresh	known concentration	amount in (gram) for preparation of 1 litre of 1% solution, (i.e. 10g per litre)	ratio of volume of chlorine product to volume of additional pure water to produce 1% solution
Sodium hypochlorite				
commercial strength	up to 15%	180g/l (14%)	71	1:13
household bleach	up to 5%	60g/l (4%)	250	1:3
Javel water	about 1%	10g/l (1%)	1000	undiluted
Chlorinated lime (Bleaching powder)	up to 35%	300g/kg (30%)	33	-
High Test Hypochlorite (HTH) powder or tablets	up to 70%	660g/kg (66%)	15	-

Household water treatment 1

particularly important during storage. Make sure that the chemicals have sufficient contact time with the pathogens (about 30 minutes in the case of chlorine) to destroy them. Chemicals left in the water will, to some extent, protect it from further contamination. To obtain the desirable residual chlorine value of between 0.3 and 0.5mg/l in the treated water, you will need to add a much higher value. This 'residual' effect cannot be obtained from solar disinfection or boiling.

Chemical disinfection is only feasible if the disinfecting chemical is readily available and affordable – and if users accept the resulting taste. As outlined above, the householder needs a reliable way of accurately adding small amounts of chemical to the raw water. Regular disinfection is necessary, as people who drink disinfected water lose their immunity to some diseases.

Solar disinfection

Ultra-violet radiation in sunlight will destroy most pathogens. You can improve the effectiveness of this process by increasing the temperature (although the temperature of the water does not need to rise much above 50°C). If clear water is exposed to sunlight, after a time it is usually rendered free of bacterial pathogens. In tropical regions, a safe exposure period is about five hours, centred around midday. One easy method of treating the water is to expose bottles of water to the sun. The SODIS system uses half-blackened bottles to increase the heat gain, with the clear side of the bottle facing the sun. For greater effectiveness place the bottle (black side down) on a corrugated-iron roof (see Wegelin and Sommer, 1998). The water can also be held in a plastic bag.

Reed (1997) has found that vigorously shaking three-quarters-full bottles to increase the oxygen content of the water to a high value before exposing it to sunlight considerably improves the effectiveness of solar disinfection. Further sporadic shaking during exposure is also beneficial. People are unlikely to want to drink the warm, treated water, so allow it to cool.

Another method of solar disinfection (called solar pasteurization) uses solar radiation to *heat* water to about 70°C for about 15 minutes to kill off the pathogens; it does not feature in this Technical Brief because it is considered too complicated for use by low-income households.

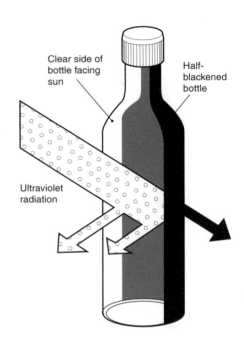

The SODIS system

Further reading

Heber, G., *Simple Methods for the Treatment of Drinking-Water*, GATE, F. Vieweg, Braunschweig, 1985.
IRC, *Community Self-Improvements in Water Supply and Sanitation: A Training and Reference Manual for Community Health Workers, Community Development Workers and other Community-Based Workers*, Training Series No. 5, IRC, The Hague, 1988.
Miller, DeWolfe, 'Boiling drinking-water: A critical look', *Waterlines*, Vol.5, No.1, IT Publications, London, 1986.
Pickford, J.A. (ed.) *The Worth of Water: Technical briefs on health, water and sanitation*, IT Publications, London, 1991.
Reed R., 'Sunshine and fresh air: A practical approach to combating water-borne disease', *Waterlines*, Vol.11, No. 4, IT Publications, London, 1997.
Wegelin M. and Sommer B., 'Solar Water Disinfection (SODIS): Destined for worldwide use?', *Waterlines*, Vol.16, No.3., IT Publications, London, 1998.

Prepared by Brian Skinner and Rod Shaw

WELL

WATER AND ENVIRONMENTAL HEALTH AT LONDON AND LOUGHBOROUGH (WELL) is a resource centre funded by the United Kingdom's Department for International Development (DFID) to promote environmental health and well-being in developing and transitional countries. It is managed by the London School of Hygiene & Tropical Medicine (LSHTM) and the Water, Engineering and Development Centre (WEDC), Loughborough University.

Phone: +44 1509 222885   Fax: +44 1509 211079   E-mail: WEDC@lboro.ac.uk   http://www.lboro.ac.uk/well/

59. Household water treatment 2

This Technical Brief is the second of two which examine the treatment of water at household level. The first (No.58) introduced the topic and covered treatment by straining, storage, settlement, solar disinfection, chemical disinfection and boiling. This Brief considers treatment by coagulation, flocculation, filtration and solar distillation and covers aspects of the reduction of some chemical concentrations.

Coagulation and flocculation

If raw water contains a large amount of fine suspended solids, coagulation and flocculation can be used to remove much of this material. In coagulation, a substance (usually in a liquid form), is added to the water to change the behaviour of the suspended particles. It causes the particles, which previously tended to repel each other, to be attracted towards each other, or towards the added material.

Coagulation takes place during a rapid mixing/stirring process which immediately follows the addition of the coagulant.

The flocculation process, which follows coagulation, usually consists of slow, gentle stirring. During flocculation, as the particles come into contact with each other they cling together to form larger particles which can be removed afterwards by settlement (see TB 58) or filtration. A chemical which is often used is alum (aluminium sulphate). Natural coagulants include some types of clay (e.g. bentonite) and powdered seeds of the *Moringa olifeira* tree. The best type of coagulant and the required dose will depend on the physical properties (particularly the alkalinity/acidity) of the raw water and the amount and type of suspended solids.

Filtration

A number of processes take place in filters, including mechanical straining; absorption of suspended matter and chemicals; and, particularly in slow sand filters, biochemical processes. Depending on the size, type and depth of filter media, and the flow rate and the physical properties of the raw water, filters can remove suspended solids, pathogens, and certain chemicals, tastes, and odours.

Straining, and settlement (both described in TB 58) are treatment methods which usefully precede filtration to reduce the amount of suspended solids which enter the filtration stage. This increases the period for which a filter can operate before it needs cleaning. Coagulation and flocculation are also useful treatments to precede settlement and improve still further the removal of solids before filtration. If an effective system of removing the flocculated particles from the filter is feasible, then coagulation and flocculation can also be used before coarse sand filtration. Larger pathogens (e.g. parasitic worm eggs) are more readily removed by filtration than smaller pathogens (e.g. viruses).

Sand filtration

Slow sand filtration

In slow sand filtration the water passes slowly (e.g. flow velocity of 0.1 to 0.2m/h – i.e. $m^3/m^2/h$) downwards, through a bed of fine sand. For the filter to perform well there should be no sudden changes in the flow rate and the water should not be very turbid (cloudy with suspended solids), or the filter will quickly become blocked. Good slow sand filters can produce good quality drinking-water.

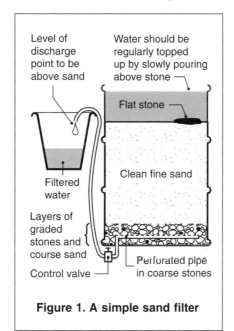

Figure 1. A simple sand filter

There are a number of processes which improve the water quality as it passes through the filter, but pathogens are mainly removed in the very top layer of the filter bed where a biological film (called the 'schmutzdecke') builds up. In a well-designed and well-operated filter this film strains out bacteria. Deeper in the sand bed bacteria that pass through the schmutzdecke are killed by other micro-organisms, or they become attached to particles of sand until they die.

The schmutzdecke takes time to become effective, so water needs to flow though a new filter for at least a week before the filter will work efficiently.

The raw water should contain a fair amount of oxygen to promote the useful biological activity both in the schmutzdecke and further down into the filter bed.

After a period of use, the material filtered out of the water blocks the surface of the sand and reduces the flow rate to an unacceptable level. When this happens the filter is drained to expose the sand, and the top 15 to 20mm of the bed is carefully removed. When the filter is restarted, it takes a few days before the schmutzdecke builds up again to provide good quality water so, during this period, the water should not be used for potable purposes.

When successive cleaning operations cause the depth of sand to reach the minimum acceptable value (conventionally 650mm thick (see TB 15)), additional clean sand needs to be added to the bed.

105

Household water treatment 2

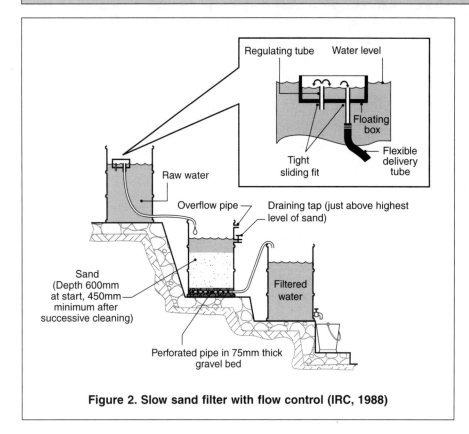

Figure 2. Slow sand filter with flow control (IRC, 1988)

When a slow sand filter is first started, water should be added from below so that it rises through the sand, pushing out the air. If water is added at the top, some air is likely to become trapped between the sand particles, adversely affecting the subsequent performance of the filter.

Many designs of household sand filters, such as that shown on the page 105, do not satisfy the recommendations for minimum depth of sand and for stable flow rate and, consequently their ability to remove all pathogens is suspect. A better, but more complicated, design of filter which uses a constant flow device is shown in Figure 2. Tests should always be carried out to check the effectiveness of any chosen filter design *before* it is promoted.

Rapid sand filtration

In this method the sand used is coarser than for slow sand filtration and the rate of flow is faster (conventionally the velocity of flow is between 4 and 8m/hr). Rapid sand filtration is used for removing suspended solids from water and is particularly effective after coagulation and flocculation. No schmutzdecke develops on these filters, so they are not effective at removing pathogens; the filtered water should subsequently be disinfected or passed through slow sand filters. There are two main types of rapid sand filter; downflow and upflow.

In a *downflow* filter the water flows down through a layer of sand, ideally between 1 and 2m deep, below a depth of water of between 1.5 and 2.5m (although these depths are rarely practical for household filters which usually have shallower depths and are therefore less effective). When this type of filter becomes clogged, the flow is reversed to mobilize the sand particles and wash out the trapped solids. The operation of this type of filter is normally too complex for household use.

In *upflow* rapid sand filters the water passes up through the sand. To clean out debris trapped in the sand, the flow is made to reverse suddenly by opening a fast-acting valve below the filter bed. To prevent the build up of deposits in the sand, backwashing may be carried out every day, although if sufficient water and pressure is available for backwashing, a longer period will be acceptable. This type of filter is sometimes used at household level with shallower depths of sand and slower flow rates than for the conventional downflow filter mentioned above. The filter illustrated in Figure 3 is one shown by Heber (1985). It uses a 300mm depth of sand and a filtration rate of 0.5 to 1.5 m/h.

Childers and Claasen (1987) and Singh and Chaudhuri (1993) give details of the Unicef upward-flow water filter which works on a slightly different principle to conventional rapid upflow filters. The Unicef filter has three beds of media, a lower bed of fine sand followed by a bed of charcoal which is covered by a bed of coarse sand. It makes use of microbial action to reduce the number of pathogens, and regular backwashing is discouraged. It is capable of treating up to 40 l/day but is not effective at removing all pathogens.

Charcoal filters

As in the case of the UNICEF filter, granular charcoal can be used during filtration. It can be quite effective at removing some tastes, odours and colour. However, there is evidence that sometimes charcoal, particularly if not regularly replaced, can become the breeding ground for some harmful bacteria.

Some disadvantages of sand filters:

- good household sand filters are not cheap to construct;
- owners of such filters need to be well motivated to operate and carefully clean the filters correctly, and periodically to carry out the time-consuming task of renewing the sand bed. If any of these tasks are not carried out properly, the quality of the water will be unreliable;
- many of the cheaper household sand-filter designs are not able to produce pathogen-free water;
- an alternative source of potable water, or sufficient stored, treated water, needs to be available during the days immediately after the cleaning or re-sanding of slow sand filters.

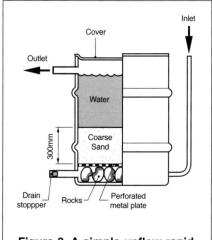

Figure 3. A simple upflow rapid sand filter (Heber, 1985)

Household water treatment 2

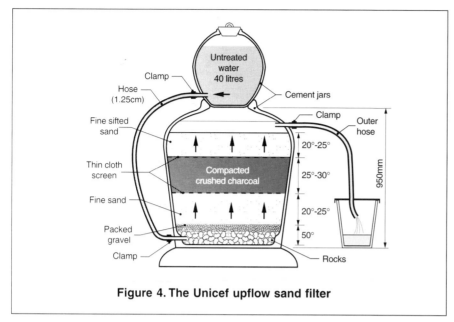

Figure 4. The Unicef upflow sand filter

Ceramic filters

The purifying element in these filters is a porous, unglazed, ceramic cylinder (often called a candle) which can be locally produced (Heber, 1985), but is usually mass-produced in factories.
Manufactured filter units like that illustrated in Figure 5 (a) are available but are costly. If filter candles are available they can be fitted to earthenware pots (b); an alternative arrangement, which avoids the need for watertight connections through the jars, is to use a siphon pipe (c); open porous-clay jars (d) can also be used. Ceramic filters are appropriate only for fairly clear water because they block quickly if the water contains suspended particles. Their effectiveness depends on the size of the pores in the clay. Filters with very small pore sizes can remove all pathogens. The impurities are deposited on the surface of the candle, so need to be regularly scrubbed off to maintain a good flow rate. Boiling the filter after it has been cleaned is also recommended to kill off the pathogens trapped in the pores, but some filters are impregnated with silver to kill micro-organisms. The scrubbing wears down the ceramic material, so periodically the candle needs to be replaced before it becomes too thin to guarantee the removal of all pathogens.

Reducing concentration of chemicals in water

Desalination

Excessive chemical salts in water make it unpalatable. Desalination by distillation (TB 40) produces water without chemical salts and various methods can be used at household level, for example to treat sea water. Desalination is also effective in removing other chemicals like fluoride, arsenic and iron. The water produced is relatively tasteless unless a little salt is added.

Iron removal

High iron content (above 0.3mg/litre), is sometimes found in groundwater collected from boreholes. It can also be a result of the corrosion of steel (e.g. pipes, borehole casings and screens) from the action of acidic groundwater. Iron precipitates cause water discoloration and can impart an unpleasant metallic taste and odour to water as well as causing the staining of food and laundry. Iron is not known to have any detrimental effects on human health, but may cause an otherwise good quality groundwater source to be rejected in favour of a bacteriologically infected surface water source. The presence of organic compounds in the water significantly increase the concentration of iron held in solution. The metabolism of some bacteria is reliant on iron and they produce a red-brown slime; decay of these bacteria also produces unpleasant odours. Treatment methods are relatively simple, being based principally on aeration followed by filtration. Many different designs of small, simple, community-level iron-removal plants have been used with handpumps, but they need commitment from someone to carry out the periodic cleaning of the stones and sand which are used for absorption and filtration. Some removal methods use biological processes (Tyrrel et al. 1988). There is little information published about household iron-removal plants. However, aeration followed by settlement, and preferably also sand filtration, is usually effective at removing excess iron.

Manganese removal

Excessive manganese (above 0.1 mg/l) causes similar staining problems to excessive iron. Some forms of manganese can be removed by aeration followed by settlement or filtration.

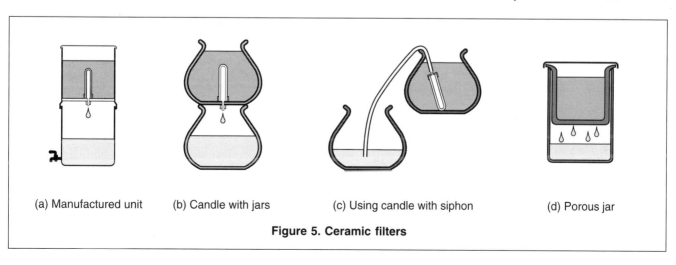

(a) Manufactured unit (b) Candle with jars (c) Using candle with siphon (d) Porous jar

Figure 5. Ceramic filters

107

Household water treatment 2

Fluoride removal
Excessive fluoride (above 1.5mg/l), which is sometimes found in groundwater, can damage bones and teeth. There are a wide variety of systems for reducing excessive fluoride and the effectiveness of each depends on various factors such as the initial concentration of fluoride, the pH of the water (a measure of the acidity or alkalinity) and the hardness of the water. The methods most suited to domestic treatment in low-income communities include lime softening or the use of pre-treated bone. One system which seems to have had considerable success in the field is the Nalgonda system which combines the use of lime (to soften the water) with alum (as a coagulant) followed by settlement; the technique is used simultaneously with chlorination to ensure disinfection of the water.

Arsenic removal
Dangerous levels of arsenic can be found naturally in groundwater and surface water, but can also result from industrial pollution. Excessive amounts are toxic to humans, resulting in various diseases including cancer. High levels are a growing problem in groundwaters in some countries like Bangladesh. The effectiveness of any treatment process depends on the specific form of arsenic found in the water and the type of coagulant and filtration material used to purify the water. Important research is presently being carried out into arsenic removal to find appropriate solutions. Treatment processes which include the addition of lime to soften the water, followed by settlement, have been in use for some time.

Combining treatment methods
Methods used to remove chemicals do not necessarily also remove pathogens. For this reason, disinfection or filtration using a ceramic filter or a well-designed slow sand filter, is likely to be necessary to produce an acceptable quality of drinking-water.

Further reading

Heber, G., *Simple methods for the treatment of drinking water,* GATE, F. Vieweg, Braunschweig, 1985.
IRC, *Community Self-Improvements in Water Supply and Sanitation: A training and reference manual for community health workers, community development workers and other community-based workers,* Training Series No. 5, International Water and Sanitation Centre (IRC), The Hague, 1988.
Hofkes E.H. (ed.), L.Huisman, B.B.Sundaresan, J.M. De Azevedo Netto and J.N. Lanoix, 'Small Community Water Supplies', Technical Paper Series No. 18, International Reference Centre for Community Water Supply and Sanitation (IRC), Wiley, Chichester, 1983.

Coagulation, flocculation
Sutherland, J.P., G.K.Folkard and W.D.Grant, 'Natural coagulants for appropriate water treatment: A novel approach', *Waterlines,* Vol. 8, No. 4 , IT Publications, London, 1990.
Dian Desa, 'Water Purification with Moringa seeds', *Waterlines,* Vol.3, No.4, IT Publications, London, 1985.
Jahn, Samia Al Azharia, *Proper use of Natural Coagulants for Rural Water Supplies: Research in the Sudan and a guide for new projects,* GTZ, Eschborn, 1986.

Filtration
Childers, L. and F. Claasen, 'Unicef's upward-flow waterfilter', *Waterlines,* Vol. 5, No. 4, IT Publications, London, 1987.
Singh,V.P. and M. Chaudhuri, 'A perfomance evaluation and modification of the UNICEF upward-flow water filter', *Waterlines,* Vol.12, No.2, IT Publications, London, 1993.

Iron removal
Ahmed F. and P.G. Smith, 'Design and performance of a community type iron removal plant for hand pump tubewells', *Journal of the Institution of Water Engineers and Scientists,* Vol.41, 1987.
Tyrrel S., Gardner S., Howsam P. and Carter R., 'Biological removal of iron from well-handpump water supplies', *Waterlines,* Vol.16, No.4, IT Publications, London, 1988.

Fluoride removal
NEERI, *Deflouridation,* National Environmental Engineering Research Institute, Nagpur, 1987.

Relevant technical briefs
(Relevant Technical Brief numbers are shown in the text preceded by TB.) 11 Rainwater Harvesting;15 & 21 Slow sand filter design; 17 & 19 Health, water and sanitation (1 & 2);40 Desalination; 46 Chlorination of community water supplies; 47 Improving pond water; 58 Household water treatment 1.

Prepared by Brian Skinner and Rod Shaw

WELL

WATER AND ENVIRONMENTAL HEALTH AT LONDON AND LOUGHBOROUGH (WELL) is a resource centre funded by the United Kingdom's Department for International Development (DFID) to promote environmental health and well-being in developing and transitional countries. It is managed by the London School of Hygiene & Tropical Medicine (LSHTM) and the Water, Engineering and Development Centre (WEDC), Loughborough University.

Phone: +44 1509 222885   Fax: +44 1509 211079   E-mail: WEDC@lboro.ac.uk   http://www.lboro.ac.uk/well/

60. Water clarification using *Moringa oleifera* seed coagulant

The removal of organic and inorganic material from raw water is essential before it can be disinfected for human consumption. In a water treatment works, this clarification stage is normally achieved by the application of chemical coagulants which change the water from a liquid to a semi-solid state. This is usually followed by flocculation, the process of gentle and continuous stirring of coagulated water, which encourages the formation of 'flocs' through the aggregation of the minute particles present in the water. Flocs can be easily removed by settling or filtration. For many communities in developing countries, however, the use of coagulation, flocculation and sedimentation is inappropriate because of the high cost and low availability of chemical coagulants, such as aluminium sulphate and ferric salts.

This Technical Brief gives an overview of the application of an indigenous, naturally derived coagulant, namely seed material from the multi-purpose tree *Moringa oleifera* Lam. *(M.oleifera)* which offers an alternative solution to the use of expensive chemical coagulants.

Background

How do the seeds work?

The seed kernels contain significant quantities of a series of low molecular-weight, water-soluble proteins which, in solution, carry an overall positive charge. The proteins are considered to act similarly to synthetic, positively charged polymer coagulants. When added to raw water the proteins bind to the predominantly negatively charged particulates that make raw waters turbid (silt, clay, bacteria etc.). Under proper agitation these bound particulates then grow in size to form the flocs, which may be left to settle by gravity or be removed by filtration.

Household water treatment

The traditional use of the *M.oleifera* seeds for domestic household water treatment is limited to rural areas in Sudan. Village women, collecting their water from the River Nile, place powdered seeds in a small cloth bag with a thread attached. This is then swirled around in the turbid water to promote coagulation and flocculation. The flocculated solids are allowed to settle and the treated water is removed before boiling and subsequent consumption.

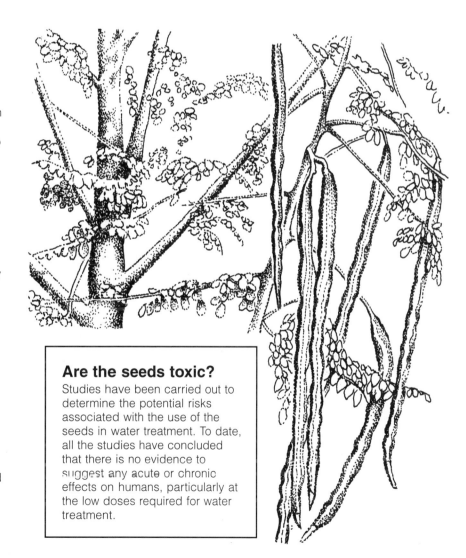

Are the seeds toxic?

Studies have been carried out to determine the potential risks associated with the use of the seeds in water treatment. To date, all the studies have concluded that there is no evidence to suggest any acute or chronic effects on humans, particularly at the low doses required for water treatment.

Water clarification using *M. oleifera* seed coagulant

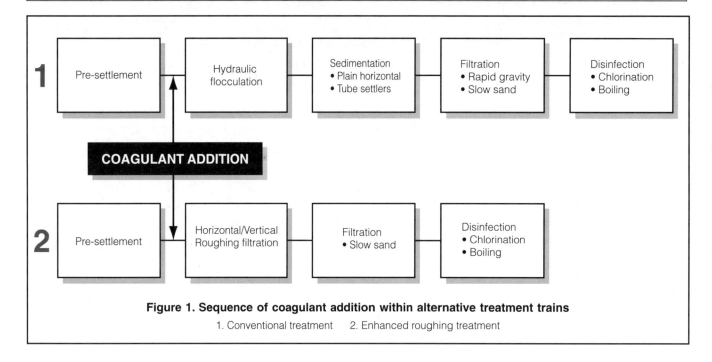

Figure 1. Sequence of coagulant addition within alternative treatment trains
1. Conventional treatment   2. Enhanced roughing treatment

Community water treatment systems

For practical reasons of solution preparation, the use of powdered seed kernels is only recommended for treatment systems up to 10m^3/hour.

As with all coagulants, the effectiveness of the seeds may vary from one raw water to another. Jar testing should be undertaken to determine their effectiveness on a particular water, and to establish preliminary dosing regimes depending on the season. The practical application of dosing solutions is exactly the same as for all other coagulants. Figure 1 (above) demonstrates the stage of application in two alternative treatment 'trains'.

Two further advantages of seed treatment are that:

- the effectiveness is, in general, independent of raw water pH; and

- the treatment does not affect the pH of the treated water.

Coagulant solution preparation

Seed solutions may be prepared from either seed kernels or the solid residue ('presscake') obtained following the extraction of seed oil.

Shelled whole seed

(1) Seed pods are allowed to mature and dry naturally to a brown colour on the tree.
(2) The seeds are removed from the harvested pods, and shelled.
(3) The seed kernels are crushed and sieved (0.8mm mesh or similar). Traditional techniques used to produce maize flour have been found to be satisfactory.
(4) The finely crushed seed powder is mixed with clean water to form a paste, and is then diluted to the required strength. Dosing solutions can be prepared from 0.5 to 5 per cent concentration (e.g. 0.5 to 5g/l).
(5) Insoluble material is filtered out using either a fine mesh screen or muslin cloth.
(6) The solution is ready for use.

Presscake

Presscake should be ground to a fine powder and sieved (0.8mm mesh or similar). Solution preparation then follows steps 4 to 6 on the left.

Note:
Solution containers should be cleaned between batches to remove insoluble seed material. Fresh solutions should be prepared every eight hours.

Crushing seed kernels and sieving

Water clarification using *M. oleifera* seed coagulant

Coagulant dose requirement

As for all coagulants, the amount of seed required will vary depending on the raw water source and on the raw water quality. One advantage of seed use is that, in general, there is a wide dose range over which effective treatment may be achieved and maintained. The dose ranges shown in Table 1 are given as a guide only, and jar testing should be carried out to determine more specific dose requirements for the raw water in question. Dosages are given as equivalent weight of seed powder or presscake material required to make up the dosing solution.

Raw water turbidity (NTU)	Dose range mg/l
< 50	10 - 50
50 - 150	30 - 100
>150	50 - 200

Table 1. Dose requirements as a function of raw water turbidity

Seed requirement

The area under cultivation to produce the annual seed requirement depends on the size of the treatment works and raw water quality (as noted in Table 1). Assuming the average seed kernel yield for a mature tree is 3kg, then at an average seed dose of 100mg/l the harvest from a single tree will treat 30 000 litres of water. Using the same assumptions and a recommended tree spacing of 3m, the harvest from 1ha of mature trees (approx. 3000kg) would treat 30 000m^3 of water. This equates to a small treat-ment works producing 10m^3 per hour if operated eight hours a day for a full year.

Moringa oleifera (M.oleifera) General information

Description

A native of northern India, *M.oleifera* is now grown widely throughout the tropics. It is sometimes known as the 'drumstick' or 'horseradish' tree. Ranging in height from 5 to 12m with an open, umbrella-shaped crown, straight trunk and corky, whitish bark, the tree produces a tuberous tap root. The evergreen or deciduous foliage (depending on climate) has leaflets 1 to 2 cm in diameter; the flowers are white or cream coloured. The fruits (pods) are initially light green, slim and tender, eventually becoming dark green, firm and up to 120cm long, depending on the variety. Fully mature, dried seeds are round or triangular, the kernel being surrounded by a lightly wooded shell with three papery wings.

Climate and soil conditions

The *M. oleifera* prefers hot, semi-arid regions (annual rainfall 250-1500mm), although it has been found to adapt well to hot, humid, wet conditions with annual rainfall in excess of 3000mm. Considered to be suitable only for lowland cultivation at altitudes less than 600m, the adaptability of the tree was demonstrated by the discovery of natural strands at altitudes of 1200m in Mexico. Although preferring well-drained sandy or loamy soils, heavier clay soils will be tolerated, although water logging should be avoided. The tree is reported to be tolerant of light frosts and can be established in slightly alkaline soils of up to pH9.

Cultivation

The tree grows rapidly from seeds or cuttings, and growth up to 4m in height; flowering and fruiting have been observed within 12 months of planting out. In areas where the climate permits, e.g. southern India, two harvests of pods are possible in a single year. Recent estimates suggest that, for a spacing of 3m, a likely annual seed yield is 3 to 5 tonnes per hectare.

Moringa, part of fruit cut vertically

Moringa flower

Seed: entire and cut vertically

Moringa fruit

Water clarification using *M. oleifera* seed coagulant

Additional products and uses of *M. oleifera*

Vegetable
- Green pods, fresh and dried leaves

Oil
- Seeds contain up to 40% of oil by weight
- Used for cooking, soap manufacture, cosmetic base and in lamps

Other uses
- All parts of the plant are used in a variety of traditional medicines
- Leaves are useful as animal fodder
- Presscake, obtained following oil extraction, is useful as a soil conditioner
- Grown as live fences and windbreaks
- Fuelwood source after coppicing (cutting back the main stem to encourage side shoots).
- As an intercrop with other crops
- Wood pulp may be used for paper-making.

The leaves have outstanding nutritional qualities, among the best of all perennial vegetables. The protein content is 27 per cent and there are also significant quantities of calcium, iron and phosphorus, as well as vitamins A, B and C. This nutritional value is particularly important in areas where food security can be threatened by periods of drought. *M. oleifera* leaves can be harvested (and dried) during dry seasons when there are no other fresh vegetables available. The immature green pods are consumed by Asian populations world-wide and canned pods are exported from India. The seeds contain up to 40 per cent by weight of oil and the fatty acid profile of the oil shows it to be, on average, 73 per cent oleic acid. The oil approaches the high quality of olive oil.

Further reading

Jahn, S.A.A., *Proper use of African natural coagulants for rural water supplies*, Manual No. 191, GTZ, Eschborn, 1986.

Jahn, S.A.A., 'Simplified water treatment technologies for rural areas', *GATE*, GTZ, Issue 1, Eschborn, 1989.

Morton, J.F., 'The horseradish tree, *Moringa pterygosperma* (Moringaceae): A boon to arid lands?' *Economic Botany* 45 (3), 1991, pp.318-33.

Schulz, C.R. and Okun, D.A., *Surface Water Treatment for Communities in Developing Countries*, Wiley / IT Publications, London, 1984.

Sutherland, J.P., Folkard, G.K., Mtawali, M.A. and Grant, W.D., '*Moringa oleifera* at pilot/full scale'. In J. Pickford et al. (ed.) *Water, Sanitation, Environment and Development: Proceedings of the 19th WEDC Conference*, WEDC, Loughborough, 1994.

Prepared by Geoff Folkard, John Sutherland and Rod Shaw

WATER AND ENVIRONMENTAL HEALTH AT LONDON AND LOUGHBOROUGH (WELL) is a resource centre funded by the United Kingdom's Department for International Development (DFID) to promote environmental health and well-being in developing and transitional countries. It is managed by the London School of Hygiene & Tropical Medicine (LSHTM) and the Water, Engineering and Development Centre (WEDC), Loughborough University.

Phone: +44 1509 222885   Fax: +44 1509 211079   E-mail: WEDC@lboro.ac.uk   http://www.lboro.ac.uk/well/

61. On-plot sanitation in urban areas

This Technical Brief examines the key issues involved in providing sanitation to low-income urban communities. We clarify differences between *on-plot* and *on-site* systems, and discuss why people lack latrines, what users want, optimum plot size, and common operational problems and maintenance issues. The findings are based on extensive consultation with urban householders in Africa and Asia.

On-plot sanitation...?

On-plot sanitation refers to types of sanitation that are contained within the plot boundaries occupied by a dwelling. Commonly, on-plot sanitation is equivalent to 'household latrine', but may also include facilities shared by several households living together on the same plot. Amongst the most commonly found on-plot sanitation technology types are:

- Unimproved pit latrines
- Lid-covered pit latrines
- Ventilated improved pit latrines
- Double-pit pour-flush latrines
- Pour-flush toilets to septic tank
- Bucket/pan latrines

By contrast, the more commonly known *on-site* sanitation includes communal facilities which are self-contained within the site, in contrast to sewerage and dry latrines where excreta is removed from the site.

Amongst some authorities and sector professionals there is an underlying feeling that whilst on-plot sanitation is appropriate for rural areas, it is generally unsuitable in the urban context, unless viewed as a (preferably short-term) route to 'better' forms of sanitation.

In practice, given the continuous growth of urban populations and the high incidence of low-income people in slums and peri-urban areas, there is no possibility of providing all urban inhabitants with sewerage. Other systems need to be employed. Well-maintained and constructed on-plot systems offer a viable alternative.

Guidelines

This technical brief presents five key questions which are central to the adoption of on-plot sanitation in urban areas, and provides specific guidelines in relation to each.

1. Why no household latrine?

Available literature emphasizes the importance of the lack of physical space in the urban environment as a key feature explaining the absence of household sanitation.

The factors which determine whether sanitation facilities are present or absent from the household plot are complex and diverse.

- A key reason is usually poverty and indebtedness, rather than lack of available space on-plot. The inability to save funds to invest in longer-term sanitation facilities, coupled with a low income, significantly restricts the choices that individuals can make.

- In cases where plot size is mentioned as a reason why a latrine has not been built, these cases are spread across a range of plot-size categories, rather than concentrating on the smallest size group.

- Plot sizes amongst households without sanitation are, on average, no smaller than those households where latrines are present.

- The relationship between cost, technology choice and income level is a complex one, which defies simple categorization. There is some evidence to suggest a relationship between unskilled employment and absence of sanitation, although this does not remain consistent for lower-cost latrine types. Similarly, skilled sources of employment are not the sole source of employment with higher-cost latrine types.

On-plot sanitation in urban areas

Choices of sanitary technology are based on a variety of factors, of which cost is just one (important) consideration.

2. Will users be satisfied with on-plot sanitation?

There is very little available work on user satisfaction as regards latrine operation in urban areas, or on changes in attitude caused by experiences with latrine operation and maintenance.

Research findings based on extensive user consultation indicate:

- In all but one technology type, users express high degrees of satisfaction with their latrine (in excess of 80 per cent recording 'satisfied' or 'very satisfied'). Bucket/pan latrines record by far the highest levels of dissatisfaction (see Table 1).

- Many users do not *perceive* there to be a problem with their latrine. Where problems are recorded, the most common include 'emptying', 'smell' and 'insects', although absolute figures are low.

- Of these three problems, 'emptying' and 'smell' have the most impact on satisfaction levels and the ability of the user to use the latrine.

3. How does plot size constrain the use of on-plot sanitation?

Critics of pit latrines often claim they are unsuitable for small plots in urban areas. In Jamaica, regulations prohibit pit latrine construction in areas where the density is higher than ten houses per acre (23 houses per hectare); in Indonesia, regulations state that areas with over 250 persons per hectare shall be classified as densely populated and shall not use on-plot excreta disposal (Alaerts et al., 1991). A manual prepared for Habitat states that the pit latrine system (except VIPs) is 'unsuitable for use in even low-density urban developments' (Roberts, 1987). The smallest plot size recommended for twin-pit pour-flush latrines in India is 26m^2 (Riberio, 1985). None of the criteria used appear to be based on reasoned argument or on evidence of performance.

- Significant proportions of households *with* sanitation facilities in working order were found on relatively small plot sizes: one third of all such cases were measured with plot areas of up to 150m^2;

Levels of user satisfaction (% of cases)					
Type	Very satisfied	Satisfied	Neither	Unsatisfied	Very unsatisfied
Bucket/pan	4	29	19	44	4
Simple pit	22	68	3	6	1
VIP	17	67	6	8	2
Pour-flush	10	73	4	8	5
WC septic-tank	22	68	3	4	3

Table 1. Levels of expressed user satisfaction by technology type

Plot sizes (m^2)					
Type	Mean	Median	Mode	Minimum	Maximum
Pour-flush	146	90	54	14	3374
Simple pit	403	306	375	28	3300
None	466	432	630	11	2700
WC septic-tank	650	576	900	27	4500
Bucket	695	600	630	70	5772
VIP	825	630	630	60	4500

Table 2. Plot-size calculations for selected technology types

just over 10 per cent on plots with an area not greater than 54m^2; plot sizes of just 14m^2 were found to have operational sanitary facilities (see Table 2).

- Levels of user satisfaction were not significantly affected by the incidence of small plot size.

- There is little indication that plot size determines technology choice. No definitive grouping or concentration of technology types was observed by recorded size categories.

- There is little indication that plot size is associated with particular operational problems. Where the most common latrine problems were noted, they were spread across all size categories.

4. What operational problems arise with on-plot sanitation?

The main problems associated with on-plot systems include odour and insect nuisance and groundwater pollution.

Odour and insect nuisance

Complaints about pit latrines most frequently mention odours and insect nuisance, yet there are few specific references to overcoming these nuisances in urban areas. Flies are a serious problem because they spread disease through feeding and breeding on faeces. Some types of mosquitoes (the *Culex* variety) breed in polluted water such as in wet latrines and may carry the disease filariasis. Controlling smells, flies and mosquitoes is, therefore, a high priority for reducing household and environmental health hazards.

In general, research findings suggest that the problem is not extensive; very few users *perceive* odour and insect nuisance to be a common problem with their latrine.

- Only 11 per cent of the total sample mention either odour (7 per cent) or insects (4 per cent) as a nuisance problem (although nuisance of this kind does have a significant impact on satisfaction levels).

On-plot sanitation in urban areas

- VIP latrines record higher than anticipated levels of odour and insect nuisance (see Tables 3 and 4). There is little conclusive evidence to suggest a link between odour and insect nuisance and the height of the vent pipe above the roof line, presence of fly screens, vent pipe colour and pipe diameter.

- Quantitative test results for insect nuisance indicate low absolute numbers of insects observed across a range of latrine types.

- Anecdotal evidence raises doubts about domestic latrines as the primary source of insect nuisance on-plot.

- Bucket/pan latrines register the highest nuisance levels of all latrine types.

Groundwater pollution

A problem that is noted in relation to on-plot sanitation is the potential for pollution of groundwater that is associated with these systems. Groundwater under or near pit latrines may become polluted, which can be a serious problem when it affects the quality of drinking-water drawn from wells and boreholes. Water in leaky pipes may also be contaminated if the pressure drops and polluted groundwater levels are above the pipes.

A particular problem in densely populated urban areas is the possible proximity of latrine pits and shallow wells on neighbouring plots. The key guideline is that a minimum distance of 15m, other than in fractured formations, between a pit and a downstream water-point, is normally sufficient to remove all contaminants.

Other critical points to note include:

- Determining the movement of viruses and bacteria in soils is extremely difficult, and involves a complex interaction of soil profile and hydraulic conductivity parameters, temperature, soil pH, and moisture-retention capacity. The clay content of the unsaturated zone is amongst the single most important indicators of the likely mobility of contaminants and their subsequent impact on groundwater pollution.

Insect nuisance (% of cases)					
Latrine type	Cases	None	Tens	Hundreds	Thousands
Bucket/pan	194	20	68	10	3
Simple pits	387	46	46	8	1
VIP	30	40	50	3	7
Pour-flush	386	71	24	5	0
WC septic-tank	127	79	21	0	0
All latrine types		54	38	6	1

Table 3. Incidence of insect nuisance by latrine type

Odour nuisance (% of cases)				
Latrine type	Cases	No smell	Slight smell	Strong smell
Bucket pan	253	10	70	20
Simple pits	388	54	37	9
VIP	48	40	54	6
Pour-flush	391	63	30	6
WC septic-tank	152	67	32	1
All latrine types		49	42	9

Table 4. User perception of the incidence of odour nuisance, by latrine type

- Larger-sized contaminants (helminths and protozoa) are normally effectively removed by physical filtration; bacteria are normally filtered by clay soils. Of most concern are water-borne viruses which are too small for even fine-grained clays to filter.

- Viruses *normally* die off within three metres of the pollution source, irrespective of soil type. Bacterial contamination is *normally* removed given sufficient depth of unsaturated soil (at least two metres) between the pollution source and water-point.

- Health risks associated with environmental pollution of groundwater must be set against the much greater hazard of open defecation, and contamination of the neighbourhood environment with excreta.

Open defecation is a serious health hazard

- If a health risk is demonstrable, investigate alternative water supplies through extending reticulation systems, as this is likely to be cheaper than centralized sewerage with treatment.

115

On-plot sanitation in urban areas

5. What happens when pits fill up?
The main guidelines relating to latrine emptying are twofold, and include advising householders that the filling/emptying cycle is likely to be between three to six years and that they need to make their own arrangements for desludging. Secondly, emptying costs are strongly location-specific; anticipated emptying costs should be investigated with local contractors during programme planning. Other findings include:

- Manual methods of emptying tend to dominate, and are especially commonplace for simple pit and pour-flush latrines. As expected, mechanical emptying tends to be associated with VIP and septic-tank latrines.

- The responsibility for emptying latrines normally falls to either the users or the contractors. Contractors play an important role in the emptying of bucket/pan and pour-flush latrines.

- Of those latrines which had been emptied, most had been used for between six and eight years. Typically, they had been emptied once or twice.

- Rates for re-filling previously emptied latrines indicate that the majority fill up after three to six years.

- Where users expressed a problem with emptying, the three most important issues were frequency, cost, and hygiene.

Summary
On-plot systems *are* appropriate for low-income urban areas, and should be considered as viable, sustainable technology choices. This research work indicates that a variety of systems are found to be working well on small plot sizes, with limited odour/insect nuisance; without significant operational problems; and to the satisfaction of the end-user. Crucially, there is a significant gulf between the perceptions of professionals and those of the community when regarding the appropriateness of on-plot sanitation in the urban context. The findings show that professionals' understanding of key issues such as insect/odour nuisance, or the operational problems associated with on-plot systems, must be advised by the opinions and perceptions of those who actually use the system.

One of the most important features of the work on on-plot sanitation is that it focuses on the *perceptions* of the users. All too often, assessments and judgements on its effectiveness and appropriateness are made from a technologically biased and purely external perspective. Many evaluations are done by those who are hardly likely themselves to be regular users of improved pit latrines. Establishing the concerns of the users of on-plot systems in urban areas and reflecting these in the guidance is a critical task.

The findings presented in this Technical Brief are drawn from Research Project R4857, *On-plot Sanitation in Low-income Urban Communities*, of the Engineering Division of the Department for International Development (DFID). This work was based on extensive consultation with urban householders (1843 cases) in three countries in Africa and Asia.

Further reading

Alaerts, G.J., Veenstra, S., Bentvelsen, M., van Duijl, L.A. et al., *Feasibility of anaerobic sewage treatment in sanitation strategies in developing countries*, IHE Report Series 20, International Institute for Hydraulic and Environmental Engineering, Delft, 1991.

Cotton, A.P., Franceys, R.W., Pickford, J.A., and Saywell, D.L., *On-plot Sanitation in Low-income Urban Communities: A review of literature.* WEDC, Loughborough University, Loughborough, 1995.

Cotton, A.P. and Saywell, D.L., *On-plot Sanitation in Low-income Urban Communities: Guidelines for selection.* WEDC, Loughborough University, Loughborough, 1998.

Mara, D., *Low-cost Urban Sanitation,* John Wiley, Chichester, 1998.

Riberio, Edgar F., *Improved Sanitation and Environmental Health Conditions: An evaluation of Sulabh International's low-cost sanitation project in Bihar,* Sulabh International, Patna, 1985.

Roberts, Martin, 'Sewage collection and disposal', *Affordable housing projects: a training manual* Prepared for United Nations Centre for Human Settlements (Habitat), Development Planning Unit, London, 1987.

Prepared by Darren Saywell and Rod Shaw

WATER AND ENVIRONMENTAL HEALTH AT LONDON AND LOUGHBOROUGH (WELL) is a resource centre funded by the United Kingdom's Department for International Development (DFID) to promote environmental health and well-being in developing and transitional countries. It is managed by the London School of Hygiene & Tropical Medicine (LSHTM) and the Water, Engineering and Development Centre (WEDC), Loughborough University.

Phone: +44 1509 222885   Fax: +44 1509 211079   E-mail: WEDC@lboro.ac.uk   http://www.lboro.ac.uk/well/

62. Emergency water supply in cold regions

During the 1990s events in the Balkans, the ex-Soviet republics, Afghanistan and Northern Iraq have demonstrated that human disasters are not limited to tropical regions of the world. In cold or mountainous regions, relief workers are faced with particular technical challenges, such as the prevention of damage to pipes and equipment caused by freezing temperatures. Following on from Technical Brief No. 44 (Emergency water supply) this Brief provides additional material for emergency and post-emergency water and sanitation staff working in regions with cool or cold climates, where freezing temperatures are likely.

General considerations for cold regions

- Initial assessment procedures should take into account climatic factors, including the possibility of seasonal freeze-ups, to determine whether cold region technology may be necessary.

- In some countries, a high level of infrastructure may have existed before the disaster. The repair of complex urban systems requires experienced engineers.

- In cold regions *'winterization studies'*, should be carried out in the summer, where possible. These should be designed to predict the possible effects a harsh winter may have on the provision of aid (see page 120).

- Winter conditions (snow and ice) can make access routes impassable. Importing water into the disaster area may not be a feasible option. Local water sources may have to be used — even if it is poor quality.

Water sources and water quality

Groundwater
In winter, groundwater is usually warmer than surface water. Using groundwater, therefore, will help to prevent water freezing in treatment systems, storage tanks and pipework. In all situations, however, levels of salinity or dissolved metals will determine whether groundwater is a suitable source or not.

Rivers and streams
Winter freezing of surface water run-off greatly reduces flow volumes and increases the concentration of ions, as more of the flow originates from groundwater sources (springs) during winter. Spring thaws lead to temporary deterioration in water quality as run-off washes impurities into the system.

Lakes
Ice is relatively pure. As surface water freezes, it rejects most salts and dissolved organic matter. These impurities, however, are concentrated in the water beneath the ice.

Emergency water supply in cold regions

Locally made water storage tanks

- Tank designs should take into account

  - the likelihood of water freezing over; and
  - the amount of damage that this causes.

- Heat lost to the air increases the likelihood of stored water freezing over. The surface area to volume ratio of the tank will affect the rate of heat loss. So:

  - a large tank will take longer to freeze over than a small one;
  - a round tank will lose heat more slowly than a rectangular one of the same volume; and
  - straight sides are better than corrugated sides as they have a smaller surface area.

- If possible, some form of insulation should be used, e.g. spray-on polyurethane foam.

- Valves can be protected by being covered and insulated where possible.

- Heat loss to the ground can cause structural instability if the frozen ground starts to thaw. Mounting the tank on an insulating concrete, or gravel, base will reduce heat transfer.

- Tank roofs should be designed to cope with extra loads arising from snow falls. Steep-angle roofs, for example, allow the snow to slide off.

- Designs should take account of rising and falling surface ice within a tank, which can cause damage to internal fittings (e.g. ladders). Internal fittings should be avoided if at all possible.

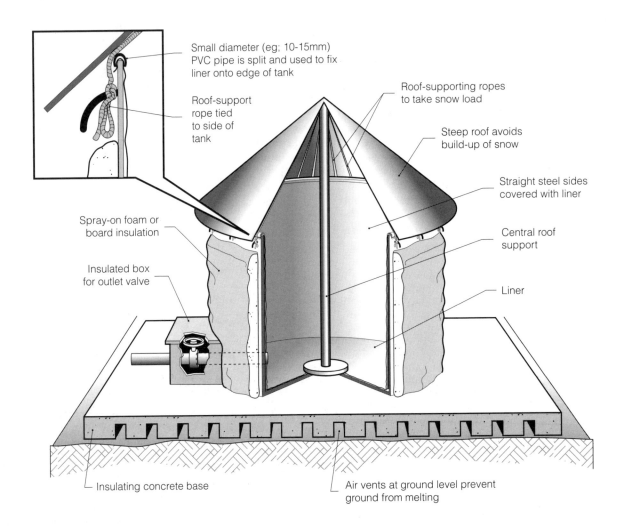

Figure 1. Temporary water storage tank, showing useful features for cold regions

Emergency water supply in cold regions

Water treatment

Low temperatures affect the rates of chemical reactions and biological processes.

Sedimentation

When treating water to remove sediments by settlement, the size of the settlement tank required can be calculated as a surface area.

Area (m²) = Design Flow Rate (m³/s)
          x Settlement Velocity (m/s)

- Jar tests are used to determine the Settlement Velocity;
- The Design Flow Rate is calculated from the size of the population.

Since settlement velocity depends on the viscosity (thickness) of the water, it is important to use water at the correct temperature. (Increased water viscosity implies a slowing of the process by a factor of 1.75 for water at 1°C compared to water at 20°C.)

Tests should be undertaken using the outside temperature to avoid underestimating the size of settlement tank required.

Slow sand filtration

The rate of flow will be slower in a cold climate both because the biological action of the 'schmutzdecke' layer is reduced and because of increased water viscocity.

Chlorination

This reaction rate is seriously affected by temperature (for every 6°C drop in temperature, the necessary contact time increases by a factor of between 1.5 and 3.5). Operators can use jar tests (for example using the Horrocks' method as described in Technical Brief No.46) to determine a suitable contact time and amount of chlorine to be added *provided that the tests are done using water samples at outside temperatures.*

Water distribution systems

The forces exerted by water expanding as it freezes and becomes ice, are the equivalent of a static head of water about 25km high! Protection of pipes and valves against frost is essential.

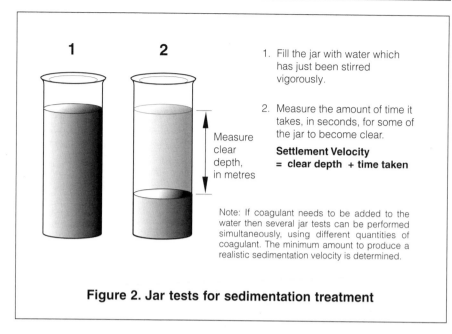

1. Fill the jar with water which has just been stirred vigorously.

2. Measure the amount of time it takes, in seconds, for some of the jar to become clear.

Settlement Velocity
= clear depth ÷ time taken

Note: If coagulant needs to be added to the water then several jar tests can be performed simultaneously, using different quantities of coagulant. The minimum amount to produce a realistic sedimentation velocity is determined.

Figure 2. Jar tests for sedimentation treatment

Immediate measures

Water is more likely to freeze if it is not moving, so:

- for a temporary supply, pipes should be *drained* when water is not flowing. For a gravity flow system, *continuous flow* can be maintained by leaving some distribution taps permanently open.

- In a pumped system, the water can be *recirculated* along dual pipe arrangements that allow water to continue flowing.

Longer-term measures

- If possible, pipes should be buried deeper than the depth of maximum frost penetration and lagged with insulation.

- Care should be taken to locate distribution points as close as possible to where people live, and preferably indoors. As well as preventing problems of taps freezing, exposure is a serious health risk, especially for the elderly, and shelter will avoid the need for people to queue in the open.

Pipe materials

Using suitable materials will reduce the probability of pipes splitting if the water freezes inside.

- Medium Density Polyethylene (MDPE) remains ductile even at very low temperatures (to -60°C).

- PVC is more brittle at low temperatures and is therefore more easily damaged than MDPE.

Pumps

Handpumps and mechanical pumps can be protected by a pump-house, which will reduce the likelihood of water freezing inside the pumps and causing damage.

Mechanical pumps

To avoid running and maintenance problems:

- Make sure that the correct grade of engine oil is used. (Oil more suitable to warmer climates may be so thick at low temperatures that it can prevent the engine from starting.)

- Use diesel suitable for cold regions. 'Gelling' can be prevented by keeping pumps indoors. (Diesel is likely to gel if transported from a warm region to a mountainous area where it is cold.)

Handpumps

- A lift pump is less likely to suffer from frost damage because the cylinder is underground.

- A small hole (approximately 3mm in diameter) cut in the riser pipe near the base will let the pump drain at night. (Note, however, that this will reduce the efficiency of the pump and limit the maximum depth from which water can be extracted).

Emergency water supply in cold regions

Winterization studies

If an emergency occurs during the summer months, in regions where winters are cold, it will be essential to carry out a winterization study. Such studies are designed to improve the efficiency of winter aid provision. They identify inadequate technology, possible logistical difficulties, and health issues that will be caused by the forthcoming winter. Measures can then be introduced to overcome potential problems before they arise.

Consider the following:

How will a harsh winter affect:

- water supply systems;
- sanitation systems;
- logistics; and
- health and motivation?

Is it worth stockpiling:

- fuel;
- food;
- materials, tools and equipment;
- suitable bags for containing wastes
- heaters, tents and blankets?

Working in cold climates

- People work more effectively in cold climates when they are warm. The value of dressing sensibly, eating enough and having regular hot drinks should not be underestimated.

- Personal kit should include warm, waterproof clothes, hat and gloves and sturdy boots. Note that the 'windchill factor' can make the climate seem much colder than the thermometer reading.

- Personal medical kits should contain adequate medication for respiratory tract infections (coughs and colds).

- Vehicles should carry shovels, snow chains and tools as well as spare tyres. Carrying food, water and four-season sleeping bags is also advisable.

- Health risks include hypothermia, snow-blindness and carbon monoxide poisoning (which can occur when small stoves are used in confined areas, with inadequate ventilation).

Further reading

Buttle, M. and Smith, M.D., *Out in the Cold: Emergency water supply and sanitation for cold regions*, WEDC, Loughborough University, Loughborough, 1999.

Davis, J. and Lambert, R., *Engineering in Emergencies*, IT Publications, London, 1995.

House, S. and Reed, R.A., *Emergency Water Sources: Guidelines for assessment and treatment*, WEDC, Loughborough University, Loughborough, 1997.

Jordan, T., *A Handbook of Gravity-flow Water Systems*, IT Publications, London, 1984.

Reed, R.A. and Shaw, R.J., 'Technical Brief No.44: Emergency Water Supply', *Waterlines*, Vol.13, No.4, IT Publications, London, 1995

Smith, D. (ed.), *Cold Regions Utilities Monograph*, ASCE, New York, 1996.

Prepared by Mark Buttle, Michael Smith and Rod Shaw

WATER AND ENVIRONMENTAL HEALTH AT LONDON AND LOUGHBOROUGH (WELL) is a resource centre funded by the United Kingdom's Department for International Development (DFID) to promote environmental health and well-being in developing and transitional countries. It is managed by the London School of Hygiene & Tropical Medicine (LSHTM) and the Water, Engineering and Development Centre (WEDC), Loughborough University.

Phone: +44 1509 222885 Fax: +44 1509 211079 E-mail: WEDC@lboro.ac.uk http://www.lboro.ac.uk/well/

63. Using human waste

Human and animal excreta has been used since ancient times as a fertilizer and soil conditioner. In Europe and North America it has been virtually replaced by artificial fertilizers, but in many other parts of the world it still plays a major role in the provision of soil nutrients.

Waste recycling is promoted for both economic and environmental reasons, but the use of fresh excreta carries considerable health hazards. This Technical Brief introduces the main issues one needs to consider to both control the process and optimize the benefits gained from using human waste, whilst minimizing the threat.

Using excreta as a resource

Human excreta is a rich source of nitrogen and other nutrients necessary for plant growth.

In many ways it is better than artificial fertilizers. It:

- encourages the formation of humus (decomposed vegetable matter) which is essential for optimum soil structure and water retention;

- contains trace elements (chemical fertilizers do not) which help protect the plant from parasites and disease;

- promotes the development of small organisms (microbes) which convert the minerals to forms that the plants can use; and

- improves the soil structure, making it easier to cultivate and to resist the effects of erosion.

The Chinese rely greatly on human excreta (sometimes known as 'nightsoil') as a fertilizer. Over 90 per cent of the quantity collected is used in agriculture.

Excreta was, and still is, recycled in a variety of ways. Table 1 lists the most common.

Urine

Urine has been used as a resource in many parts of the world for centuries.

When diluted with water it makes an excellent liquid fertilizer. Until quite recently it was used in Europe for household cleaning, softening wool, hardening steel, tanning leather and dyeing cloth. The Greeks and Romans used it for colouring their hair. African farmers use it for fermenting plants to produce dyes, and the Chinese pharmaceutical industry uses it to make blood coagulants.

Health risks

The transmission of disease through the use of untreated excreta can be widespread. Parasitic eggs and cysts may remain viable in soil, water and on plants for many months, creating a hazard for farmers and customers alike.

The primary aim of a sanitation project is to prevent the spread of disease. Where excreta is to be used as a resource, therefore, it is essential to introduce methods for protecting the users from the diseases it contains. Some of the main features of compost latrines designed to do this are described overleaf.

Table 1. Common excreta-recycling practices

Practice	Countries where used
Soil fertilization with untreated excreta	China, Japan, Korea, Taiwan, Thailand
Composted excreta used in agriculture	China, India
Excreta fed to animals	China, India, Melanesia, Nigeria, Sri Lanka
Compost latrines	Central America, Vietnam, Europe
Biogas production	China, India, Korea,
Fish food	China, India, Korea, Malaysia, Indonesia
Aquatic weed production	Vietnam, South-east Asia

Using human waste

Compost latrines

Composting is the process of biologically breaking down solid organic matter to produce a substance (compost) which is valuable as a fertilizer and soil conditioner. Excreta is mixed with other matter to reduce the moisture content, adjust the chemical balance of the mass and improve the texture.

Given a large enough organic mass and the presence of oxygen, the breakdown process will release energy. This energy will be partly used by the bacteria to reproduce, and will be partly converted to heat. If sufficient heat is generated, the temperature of the mass can rise enough to kill pathogens. The wastes generated by a single family are not enough to support such a large increase in temperature and, therefore, the process cannot be relied upon to destroy pathogenic organisms, particularly parasitic worms and cysts. Any latrine based on this process must contain other features to prevent the spread of disease. Decomposition can also take place in the absence of oxygen (anaerobically). It is slower and does not produce any rise in temperature. It may also produce a strong odour.

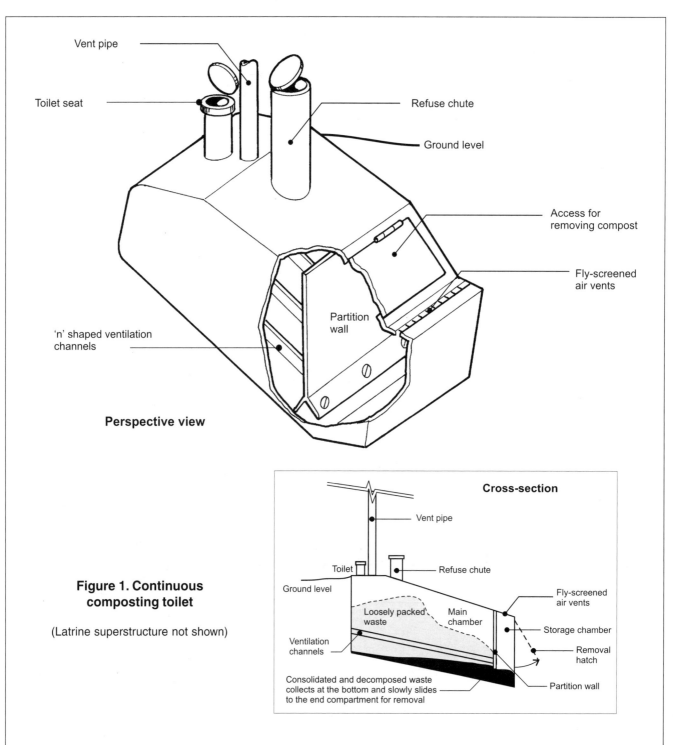

Figure 1. Continuous composting toilet

(Latrine superstructure not shown)

Using human waste

Aerobic latrines
Aerobic composting latrines must:

- keep the waste material open and relatively dry so that air can circulate freely;

- separate new wastes from old; and

- contain the wastes for long enough for any pathogenic organisms to be destroyed.

The most common design is based on a unit developed in Sweden, called the 'Multrum'.

Excreta and other wastes are added to a 3m-long receptacle which slopes away from the inlet (Figure 1). The latrine is fitted with a suspended floor made of channels shaped like an 'n' which draw air in through the storage chamber, through the decomposing mass and out via a ventilation pipe. The mass gradually slides down the suspended floor as it decomposes. Eventually it collects in the storage chamber from which it can be removed. The main chamber holds the material for about a year to ensure that the pathogens have died. The compost moisture content and chemical balance are controlled by adding vegetable waste and sawdust or ash. The process is continuous, with the weight of new material helping to push the decomposing compost towards the storage chamber.

The latrines have proved successful in small communities in industrialized countries, but it is usually necessary to install a fan on the ventilation pipe to increase ventilation, thereby controlling odours and flies.

Attempts to introduce them in other parts of the world have failed. Problems were experienced with over-use, too high a moisture content, and failure to add enough vegetable waste. These problems led to foul-smelling, unpleasant latrines which were shunned by users.

The main problem seems to have been that the users did not consider the end-product worth the effort. Family composting latrines are expensive to build, and produce only small quantities of compost. Many people prefer to purchase chemical fertilizers (which are often subsidized by the State) rather than work with human excreta.

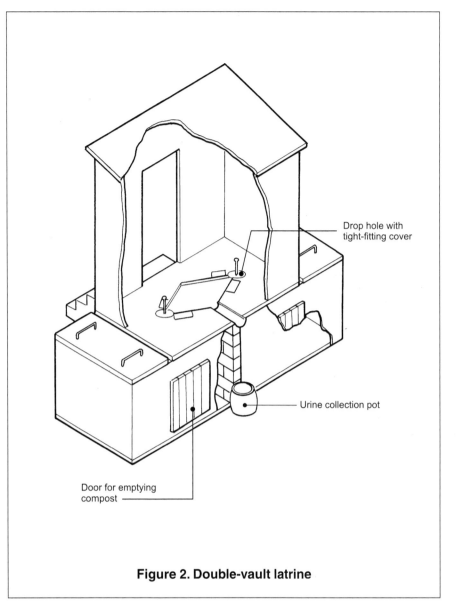

Figure 2. Double-vault latrine

Anaerobic latrines
It is easier to produce anaerobic (without air) than aerobic (with air) conditions. A moist mass of excreta and other wastes will naturally compact, which will in turn exclude the air and turn the mass 'anaerobic'. Anaerobic latrines usually work on a batch system. Excreta, vegetable waste (to control the chemical balance) and ash or sawdust (to control the moisture content) are deposited in a sealed container (Figure 2). Urine is collected separately. When the container is full it is sealed and another container can be used. When the second container is full the first container can be emptied and re-used.

There is some dispute as to how long the compost should be stored before it is used. Some people believe that it should be stored for at least two years while others recommend as little as 10 months. The longer the wastes are stored, the more pathogens will be destroyed.

Batch-composting toilets have had mixed success. Failure has generally been the result of poor understanding of the process or lack of interest in the final product. Controlling the moisture content has often caused problems, particularly in communities where water is used for anal cleaning or people bathe in the toilet cubicle.

Using human waste

Biogas

If human excreta is combined with animal and agricultural wastes, and water, it will give off gas as it decomposes. Given the right temperature and mix of wastes, much of the gas will be methane, which is flammable. The mix of gases produced is called 'biogas'.

Biogas plants have been incorporated into domestic latrines in a number of countries with mixed success. The plants are used widely in China where the gas produced is used for cooking and lighting. Plants similar to the one shown in Figure 3 take excreta and pig wastes mixed with water.

Biogas plants typically store the wastes for about 30 days. This removes some of the pathogenic organisms but by no means all. It is better to store the excreta for a period prior to or after putting them in the biogas tank.

Biogas plants can be expensive to build and difficult to operate. Poor maintenance leads to loss of gas production and blockage of the digester tank with solids. They are only appropriate in communities with a commitment to recycling organic wastes and where there are few alternative power sources.

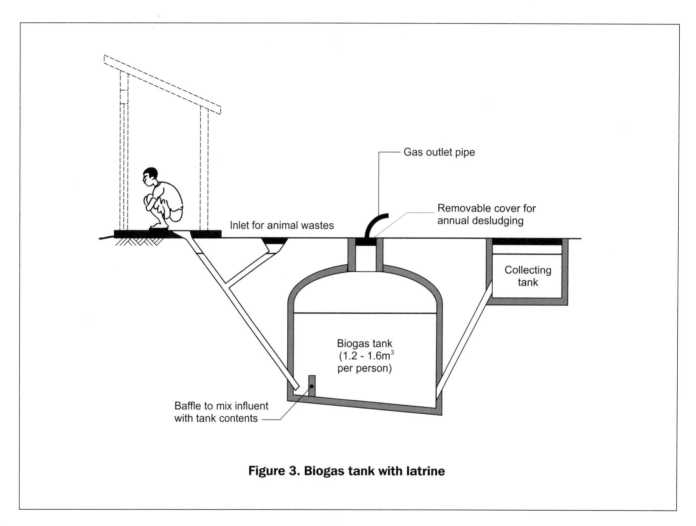

Figure 3. Biogas tank with latrine

Further information

Franceys, R., Pickford, J. and Reed, R., *A Guide to the Development of On-site Sanitation,* WHO, Geneva, 1992.
Pickford, John., *Low-cost Sanitation: A survey of practical experience,* IT Publications, London, 1995.

Prepared by Bob Reed and Rod Shaw

WELL — WATER AND ENVIRONMENTAL HEALTH AT LONDON AND LOUGHBOROUGH (WELL) is a resource centre funded by the United Kingdom's Department for International Development (DFID) to promote environmental health and well-being in developing and transitional countries. It is managed by the London School of Hygiene & Tropical Medicine (LSHTM) and the Water, Engineering and Development Centre (WEDC), Loughborough University.

Phone: +44 1509 222885   Fax: +44 1509 211079   E-mail: WEDC@lboro.ac.uk   http://www.lboro.ac.uk/well/

64. Wastewater treatment options

This Technical Brief reviews some of the options for wastewater treatment in low- and middle-income communities. It should be used as a guide to the main options available.

Wastewater management is a costly business. Once wastewaters (taken here to mean any combination of domestic sewage and industrial effluents) are produced and collected in sewerage systems, then treatment becomes a necessity. It is important to note that reducing the volume of wastewater produced and/or avoiding the need for sewerage and treatment in the first instance has many advantages; the decision to move away from properly implemented on-site sanitation should not be taken lightly.

On-site versus off-site sanitation?

On-site sanitation is often (and should be) the first option when considering a sanitation intervention. Such systems have very distinct advantages, not least because they are individual systems, so the disposal of faecal material is dispersed over a wide area, and not centralized as with a conventional sewage treatment works. One of the main disadvantages with centralized facilities is that when they go wrong, the resulting problems can be very acute.

From a *health* point of view, there is not much difference between any of the different options for sanitation (either on- or off-site) — so long as they are all functioning properly. It is largely a question of *convenience*; an off-site system where wastes are flushed off the owner's property is more convenient as it gets rid of the problem from the owner's property. Off-site sanitation is usually much more expensive than on-site.

There are instances, however, where off-site sanitation is deemed necessary — because of unsuitable ground or housing conditions for on-site systems, or because of a community's commitment to an off-site system. There is a certain amount of prestige in having an off-site connection; peer pressure is often a significant motivating force. Once the decision has been made to implement an off-site system, sewers become a necessity. Water has a large dispersion, dilution and carriage capacity, and is, therefore, used as the carriage medium in most sewer systems. Usually, potable water is supplied to the house and is used for flushing toilets, and as much as 40 per cent of household water use may be used for this purpose. Some countries do use dual supply systems where non-potable water (often sea water) is used for toilet flushing, but such a system requires more infrastructure and has obvious capital cost implications. Therefore, most sewer systems are heavy users of precious potable water supplies, which should be a factor when considering their implementation, especially in water-poor areas.

Re-use, recovery

Traditionally, sewage has been seen as a *problem* requiring treatment and disposal. Most conventional sewage treatment options are based on approaches to Northern countries' problems, which has usually meant a reduction in biodegradable organic material and suspended solids, plus perhaps some nutrients (nitrogen and phosphorous). Treatment has involved the 'removal' of these pollutants, but removal is usually conversion to another product, usually sludge. The disposal of sewage sludge is a major consideration in many locations, and it is often seen as an offensive product which is either dumped or burned.

The priorities in developing countries are often different from those in developed countries. Often the main issue is how to control pathogenic material, and any form of sanitation (on or off-site) should have this as its main objective. There are treatment options

> **NOTE**
>
> Not all bacteria are harmful!
>
> Bacteria may be:
> *Harmful,*
> *Harmless (benign)*
> *Helpful or Useful*
>
> Wastewater treatment tries to reduce the numbers of harmful bacteria
>
> Wastewater treatment encourages useful bacteria to treat wastewater

which can remove pathogenic material, notably waste-stabilization ponds.

Increasingly, sewage is being seen as a *resource*. The water and nutrient content, in particular, can be very useful for agricultural purposes (for example, through irrigation) if the sewage is treated to a suitable standard. There are treatment options which seek to use this resource potential. Traditional sewage treatment practices in South-east Asia, for example, seek to use wastes generated through pond systems which are used to cultivate fish and generate feed for animals. Some community-based approaches (in Latin America in particular) seek to separate 'grey' wastewater (non-faecally contaminated wastewater) from 'black' (faecally contaminated) water so that they can both be recycled and re-used as appropriate. In principle, the grey water can be re-used as irrigation water, and the black water/waste treated and re-used as fertilizer.

Wastewater treatment options

Sewage treatment options may be classified into groups of processes according to the function they perform and their complexity:

Preliminary: this includes simple processes such as screening (usually by bar screens) and grit removal. (through constant velocity channels) to remove the gross solid pollution.

Primary: usually plain sedimentation; simple settlement of the solid material in sewage can reduce the polluting load by significant amounts.

Secondary: for further treatment and removal of common pollutants, usually by a biological process.

Tertiary: usually for removal of specific pollutants e.g. nitrogen or phosphorous, or specific industrial pollutants.

Traditionally, sewage treatment has taken place through the implementation of large centralized schemes. Many of these do not work — and when they do not work, the resultant pollution and health problems are often severe. The reason for failure is frequently that the options chosen in the first place, are not sustainable. Often, sewage treatment is a low priority when compared to water supply, and municipal councils simply do not have the resources to keep the facilities operational. In such circumstances, there is a growing body of opinion that advocates moves towards *decentralized,* local systems, which, it is argued, could be supported by community-based organizations. Such approaches have been implemented in parts of South America.

Wastewater treatment options

Very few sewage-treatment facilities in most developing countries work. This is often because most technologies for sewage treatment are big, centralized schemes which have been developed in the North where adequate financial, material and human resources are available. Transferring these technologies to tropical low- and middle-income communities has many potential difficulties. However, there are some sewage-treatment options which are more appropriate to developing country scenarios. Such systems should generally be low-cost, have low operation and maintenance requirements, and, should maximize the utilization of the potential resources (principally, irrigation water and nutrients).

Preliminary and primary treatment are common to most sewage-treatment works, and are effective in removing much of the pollution. There are many different types of secondary process. The most common are described in the table opposite, with brief comments on their suitability for low- and middle-income countries. Tertiary treatment processes are generally specialized processes which are beyond the need of most communities.

Options for low- and middle-income communities

Most wastewater treatment processes have been developed in temperate, Northern climates. Applying them in most developing countries will have three main disadvantages:

- high energy requirements;
- high operation and maintenance requirements, including production of large volumes of sludge (solid waste material);
- they are geared towards environmental protection rather than human health protection — for example, most conventional wastewater treatment works do not significantly reduce the content of pathogenic material in the wastewater.

Aerobic versus anaerobic treatment

Most conventional wastewater treatment processes are 'aerobic' — the bacteria used to break down the waste products take in oxygen to perform their function. This results in the high energy requirement (oxygen has to be supplied) and a large volume of waste bacteria ('sludge') is produced. This makes the processes complicated to control, and costly.

The bacteria in 'anaerobic' processes do not use oxygen. Excluding oxygen is easy, and the energy requirements and sludge production is much less than for aerobic processes — making the processes cheaper and simpler. Also, the temperature in which the bacteria like to work is easy to maintain in hot climates.

However, the main disadvantages of anaerobic processes are that they are much slower than aerobic processes and are only good at removing the organic waste (the 'simple' waste, the sugary material) and not any other sort of pollution — such as nutrients, or pathogens. Anaerobic processes generally like 'steady' effluents — they are not good with coping with variations in flow or composition. For example, anaerobic processes cannot cope with shock loads of heavy metals (from industrial processes, for example).

The requirement in most low-income countries is for a low-cost, low-maintenance sewage treatment system. *Waste stabilization ponds* (WSPs) provide the best option in most cases — good levels of treatment at low capital and particularly low O&M cost. In addition, it is one of the few processes which provides good treatment of pathogenic material. This has significant application potential for re-use of the treated effluent in irrigation. The major disadvantage is that significant areas of land are needed for treatment. WSPs are used in many locations worldwide, including Africa and Asia.

Conclusion

Any wastewater treatment plant needs significant investment and O&M and control, and therefore *any* decision to implement such a facility should be carefully considered. WSPs provide the best option for a low-cost, low-maintenance system which is most effective in removing the pollutants of major concern.

Wastewater treatment options

Common options for secondary sewage treatment
(* indicates processes more suitable for developing countries)

Treatment process	Description	Key features
Activated sludge process (ASP)	Oxygen is mechanically supplied to bacteria which feed on organic material and provide treatment.	Sophisticated process with many mechanical and electrical parts, which also needs careful operator control. Produces large quantities of sludge for disposal, but provides high degree of treatment (when working well).
Aerated lagoons	Like WSPs but with mechanical aeration	Not very common; oxygen requirement mostly from aeration and hence more complicated and higher O&M costs.
*Land treatment (soil aquifer treatment – SAT)	Sewage is supplied in controlled conditions to the soil	Soil matrix has quite a high capacity for treatment of normal domestic sewage, as long as capacity is not exceeded. Some pollutants, such as phosphorus, are not easily removed.
Oxidation ditch	Oval-shaped channel with aeration provided	Requires more power than WSP but less land, and is easier to control than processes such as ASP (see below).
*Reed (or constructed wet lands) beds	Sewage flows through an area of reeds	Treatment is by action of soil matrix and, particularly, the soil/root interface of the plants. Requires significant land area, but no oxygenation requirement.
Rotating biological contractor (or biodisk)	Series of thin vertical plates which provide surface area for bacteria to grow	Plates are exposed to air and then the sewage by rotating with about 30 per cent immersion in sewage. Treatment is by conventional aerobic process. Used in small-scale applications in Europe.
Trickling (or 'percolating') filters	Sewage passes down through a loose bed of stones, and the bacteria on the surface of the stones treats the sewage	An aerobic process in which bacteria take oxygen from the atmosphere (no external mechanical aeration). Has moving parts, which often break down in developing country locations.
*Upflow anaerobic sludge blanket (UASB)	Anaerobic process using blanket of bacteria to absorb polluting load	Suited to hot climates. Produces little sludge, no oxygen requirement or power requirement, but produces a poorer quality effluent than processes such as ASP. (NOTE: other anaerobic processes exist, but UASB is the most common at present).
Waste-stabilization ponds (WSP) ('lagoons' or 'oxidation ponds')	Large surface - area ponds	Treatment is essentially by action of sunlight, encouraging algal growth which provides the oxygen requirement for bacteria to oxidize the organic waste. Requires significant land area, but one of the few processes which is effective at treating pathogenic material. Natural process with no power/oxygen requirement. Often used to provide water of sufficient quality for irrigation, and very suited to hot, sunny climates.

Wastewater treatment options

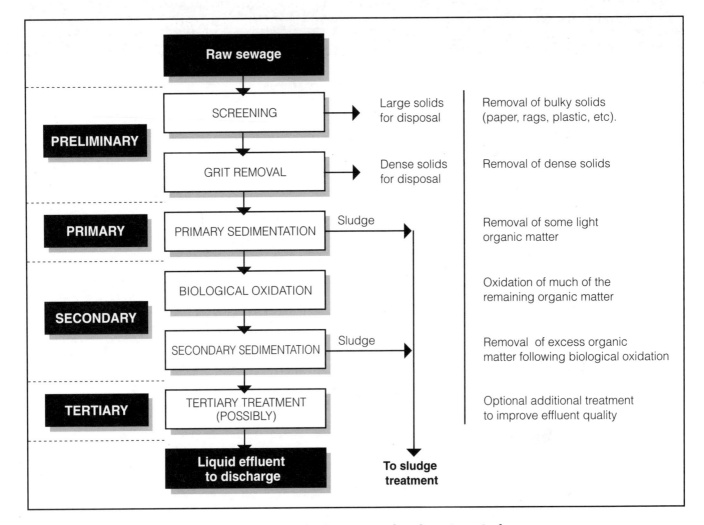

Figure 1. Typical stages in the conventional treatment of sewage

Further reading

Haandel, Adrianus C. van, and Lettinga, Gatze, *Anaerobic Sewage Treatment: A practical guide for regions with a hot climate,* John Wiley, Chichester, 1994.

Mara D.D. et al., *Waste Stabilization Ponds: A design manual for eastern Africa,* Lagoon Technology International, Leeds, 1992.

Metcalf and Eddy Inc., *Wastewater Engineering: Treatment, disposal and re-use,* 3rd edition revised by George Tchobanoglous and Franklin L. Burton, McGraw Hill Inc. International, 1991.

WPCF, *Natural Systems for Wastewater Treatment: Manual of practice,* Water Pollution Control Federation, Alexandria VA, 1990.

Prepared by Jeremy Parr, Michael Smith and Rod Shaw

WATER AND ENVIRONMENTAL HEALTH AT LONDON AND LOUGHBOROUGH (WELL) is a resource centre funded by the United Kingdom's Department for International Development (DFID) to promote environmental health and well-being in developing and transitional countries. It is managed by the London School of Hygiene & Tropical Medicine (LSHTM) and the Water, Engineering and Development Centre (WEDC), Loughborough University.

Phone: +44 1509 222885 Fax: +44 1509 211079 E-mail: WEDC@lboro.ac.uk http://www.lboro.ac.uk/well/